Journal of Applied Logics - IfCoLog Journal of Logics and their Applications

Volume 13, Number 2

April 2026

Disclaimer

Statements of fact and opinion in the articles in Journal of Applied Logics - IfCoLog Journal of Logics and their Applications (JALs-FLAP) are those of the respective authors and contributors and not of the JALs-FLAP. Neither College Publications nor the JALs-FLAP make any representation, express or implied, in respect of the accuracy of the material in this journal and cannot accept any legal responsibility or liability for any errors or omissions that may be made. The reader should make his/her own evaluation as to the appropriateness or otherwise of any experimental technique described.

ISBN 978-1-84890-508-5
ISSN (E) 2631-9829
ISSN (P) 2631-9810

College Publications
Scientific Director: Dov Gabbay
Managing Director: Jane Spurr

http://www.collegepublications.co.uk

Editorial Board

Scope and Submissions

This journal considers submission in all areas of pure and applied logic, including:

- pure logical systems
- proof theory
- constructive logic
- categorical logic
- modal and temporal logic
- model theory
- recursion theory
- type theory
- nominal theory
- nonclassical logics
- nonmonotonic logic
- numerical and uncertainty reasoning
- logic and AI
- foundations of logic programming
- belief change/revision
- systems of knowledge and belief
- logics and semantics of programming
- specification and verification
- agent theory
- databases
- dynamic logic
- quantum logic
- algebraic logic
- logic and cognition
- probabilistic logic
- logic and networks
- neuro-logical systems
- complexity
- argumentation theory
- logic and computation
- logic and language
- logic engineering
- knowledge-based systems
- automated reasoning
- knowledge representation
- logic in hardware and VLSI
- natural language
- concurrent computation
- planning

This journal will also consider papers on the application of logic in other subject areas: philosophy, cognitive science, physics etc. provided they have some formal content.

Submissions should be sent to Jane Spurr (jane@janespurr.net) as a pdf file, preferably compiled in LaTeX using the IFCoLog class file.

CONTENTS

A Representation of Explicit Knowledge and Epistemic Indistinguishability in a Logic of Awareness

Yudai Kubono
Graduate School of Science and Technology, Shizuoka University, Ohya, Shizuoka 422-8529, Japan
yudai.kubono@gmail.com

Satoshi Tojo
Department of Data Science, Asia University, Sakai, Tokyo 180-8629, Japan
tojo_satoshi@asia-u.ac.jp

Abstract

The logic of awareness, first proposed by Fagin and Halpern, addressed the problem of logical omniscience by introducing the notion of awareness and distinguishing explicit knowledge from implicit knowledge. In their framework, explicit knowledge was defined as the conjunction of implicit knowledge and awareness, each of which was represented by modal operators. Their definition, however, may derive undesirable propositions that cannot be considered explicit knowledge when Modus Ponens is applied within implicit knowledge. Hence, focusing on indistinguishability among possible worlds, dependent on awareness, we refine the definition of explicit knowledge. In our semantics, we require that the aware implicit knowledge is not necessarily explicit knowledge, though explicit knowledge must be aware as well as implicit. We employ an example of elementary geometry, where different students may or may not reach the final answer, depending on whether they are aware of learned mathematical facts. Thereafter, we formally present the syntax and the semantics of our language, named Awareness-Based Indistinguishability Logic (AIL). We prove that AIL has more expressive power than the logic of Fagin and Halpern, and show that the latter is embeddable in AIL. Furthermore, we provide an axiomatic system of AIL and prove its soundness and completeness.

Keywords: Awareness, Awareness logic, Explicit knowledge, Epistemic logic.

We thank the anonymous reviewer, Nobu-Yuki Suzuki, and Hiromi Inada for their helpful comments. This work was financially supported by 23K21869.

1 Introduction

Epistemic Logic (henceforth, EL), as a variant of modal logic, employs the modal operator K, where $K\varphi$ means "an agent knows φ." However, this K operator causes the well-known problem of *logical omniscience*. For instance, the axioms and inference rules of EL derive that an agent has knowledge of all tautologies and that when "an agent knows φ" and "the agent knows $\varphi \to \psi$," the agent always knows ψ. Many approaches have been proposed to close this gap [8]. The logic of awareness (henceforth, FH logic[1]) [7] addressed the issue of logical omniscience, incorporating the notion of *awareness* to distinguish *implicit/explicit* knowledge, where the implicit knowledge behaves as the usual K in EL while the explicit knowledge is the *aware* implicit knowledge. Even if φ is a tautology, if it is not aware by the agent, it does not belong to her explicit knowledge. This concise and intuitive idea has become a mainstay in the literature on logics of awareness [22].

Their definition, however, may derive undesirable propositions that cannot be considered explicit knowledge, when *Modus Ponens*[2] (MP) is applied within implicit knowledge. Let I, A, and E represent the operators for implicit knowledge, awareness, and explicit knowledge, respectively. We primarily focus on a valid formula in FH logic, $I(\varphi) \wedge I(\varphi \to \psi) \to (A\psi \to E\psi)$, which is entailed from (1): logical omniscience for implicit knowledge $I(\varphi) \wedge I(\varphi \to \psi) \to I\psi$ and (2): the conjunctive definition of explicit knowledge $I\psi \wedge A\psi \leftrightarrow E\psi$. This formula states that the inferred consequent is classified as explicit knowledge whenever the agent is aware of it, even if not aware of the antecedents themselves. On the other hand, one may take the view that when an agent explicitly claims to know a proposition that is derived through inference, it must be derived solely from propositions of which the agent is aware [5]. On this view, implicit knowledge conjunct with awareness need not suffice for explicit knowledge. Hence, this paper revisits the conjunctive definition of explicit knowledge, refining its representation so that $E\varphi \to I\varphi \wedge A\varphi$ but $I\varphi \wedge A\varphi \not\to E\varphi$. The following section highlights the existing gap between the formal representation of the knowledge and its intended interpretation.

Traditionally, there has been a distinction between *awareness of* and *awareness that*; the former restricts the *aware* atomic propositions and is dealt in economics [17, 15], philosophy [13, 9], and computer science [1, 21], while the latter concerns complex formulae [13, 10, 9] and furthermore the computable truth with time restriction [7, 26]. In this paper, we focus only on *awareness of*, to elucidate the revision from the preceding works.

[1]The logical semantics proposed in [7], an *awareness structure*, can be given some restrictions, and we can consider several classes of the structures and the corresponding logics.

[2]From φ and $\varphi \to \psi$, ψ is concluded.

The rest of this paper is structured as follows. In Section 2, we provide a motivating example and propose a semantic notion to refine the FH logic's definition of explicit knowledge. In Section 3, we formally define the syntax and semantics of our language, introducing a logic named *Awareness-Based Indistinguishability Logic* ($\mathcal{AIL}$). We use $\mathcal{AIL}$ to refer to the set of formulae semantically characterized, i.e., the set of all valid formulae in the class of our models. Thereafter, we demonstrate that our formalization faithfully captures the knowledge in the example introduced in Section 2. In Section 4, we prove that our logic is more expressive than an FH logic, which means that for every formula of the FH logic, there exists a formula in the language of $\mathcal{AIL}$ that has the same meaning, and show that the latter is embeddable into $\mathcal{AIL}$. In Section 5, we provide an axiomatic system **AIL** in the Hilbert style and prove its soundness and completeness. Section 6 is devoted to related work. Finally, Section 7 concludes with directions for future research.

2 Examples with Awareness and Indistinguishability

Before introducing a motivating example, we clarify the notions of awareness and knowledge focused on in this paper.

Awareness This notion concerns which information is available to the agent at a given moment and not the acknowledgement of its truth or falsity; thus, "to be aware of a proposition" means "to be able to refer to a proposition" or "to entertain a proposition."

Implicit Knowledge This comprises what an agent has independently of the agent's current awareness. Furthermore, the knowledge closes under logical consequences and includes all tautologies, thus constituting the ideal knowledge of a logically omniscient agent.

Explicit Knowledge This comprises what an agent has in the agent's awareness and hence is available for reasoning or decision-making at that moment. This knowledge includes the aspect of availability expressed by the notion of awareness, referring to what the agent actually knows at a given time.

The following example succinctly shows the difference between those notions of knowledge:

Example 1. *An agent knows the color of her smartphone, but it is not at the forefront of her mind, say, while she is reading a cookbook.*

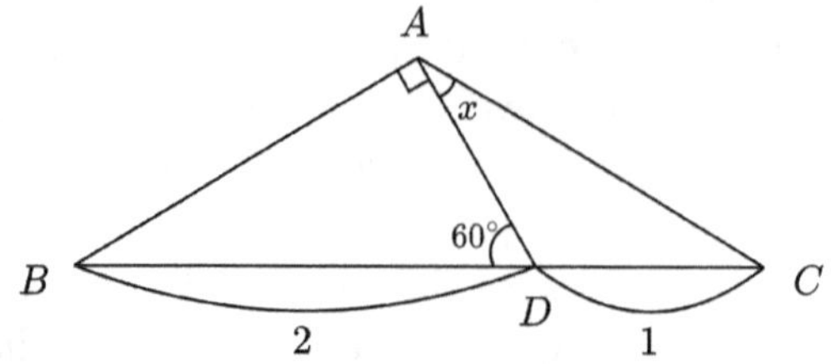

Figure 1: An elementary geometry problem.

The implicit knowledge becomes available to her when attention is drawn to it, for instance, when someone specifically asks about it.

Since the definition of awareness is given as a subset of propositions in a possible world, the explicit knowledge is only a limitation or a filtered one that originally resides in each world. Therefore, Examples 2 and 3 are inadequate in this context.

Example 2. *One day, you are introduced to a cousin who has never met before. Furthermore, you have not known that you have had such a relative.*

Although it is true that he or she is the cousin of "you", this cannot be implicit knowledge since the information does not reside in each world.

Example 3. *You are playing a chess game, fully familiar with the rules, and there is an action to checkmate; however, you do not notice it.*

In this case, there may be hidden propositions of which "you" are not aware; however, even though these atomic propositions are externalized, "you" may not be able to find the path to checkmate because of so large a search space. This raises issues of resource-boundedness tied to awareness and, in turn, to explicit knowledge. The notion of awareness discussed in this paper does not aim to capture this kind of case, and we do not pursue such resource-boundedness here. For studies addressing reasoning on explicit belief/knowledge, see [2, 9]. Nor do we address this kind of dynamics.

2.1 A Motivating Example

We introduce an elementary geometry problem as a motivating example. The problem illustrated in Figure 1 can be solved by utilizing mathematical facts, such as those related to the ratios in a 30°-60°-90° triangle and the measure of an exterior angle. Specifically, let f_i be the mathematical facts, that is:

$$\cfrac{\cfrac{f_1 \qquad k_1\colon \triangle BDA \text{ is } 30^\circ\text{-}60^\circ\text{-}90^\circ \text{ triangle} \qquad k_2\colon BD = 2}{p_1\colon DA = 1} \qquad k_3\colon DC = 1}{p_2\colon DA = DC}$$

$$\cfrac{\cfrac{f_2 \qquad k_1}{p_3\colon \angle DAC + \angle DCA = 60^\circ}\,(\dagger) \qquad p_2 \qquad f_3}{p_4\colon \angle DAC = 30^\circ}\,(\ddagger)$$

Figure 2: A proof of Figure 1. Note that the given hypotheses or derived facts are shown in the upper row of each horizontal line, and their immediate consequence is shown in its lower row, though the proof format is not based on a known formal system.

- f_1: A triangle with angles $30^\circ, 60^\circ, 90^\circ$ has sides in the ratio $1\colon \sqrt{3}\colon 2$ (opposite $30^\circ, 60^\circ, 90^\circ$, respectively);
- f_2: The measure of an exterior angle is equal to the sum of the measures of the two remote (non-adjacent) interior angles of the triangle;
- f_3: The measures of angles opposite the equal sides are equal,

k_i given knowledge for $i \in \{1, 2, 3\}$, and p_i obtained knowledge for $i \in \{1, 2, 3, 4\}$; then the proof in Figure 2 can be constructed.

We consider two students who differ in awareness and whether each can correctly respond to the question[3]: "Is angle x 30 degrees (p_4)?"

Example 4. *Two students, a and b, take a geometry exam and answer the question: "Is angle x 30 degrees?" We assume that their implicit knowledge includes f_i*[4] *and further the implications (e.g., $(f_1 \wedge k_1 \wedge k_2) \rightarrow p_1$) connecting the upper propositions to the lower one at each inference step in Figure 2. Student a is aware of each f_i, whereas b is aware only of f_1 and f_3. Since a is aware of all the mathematical facts, she knows p_4 explicitly. In contrast, b is unaware of f_2 and thus unable to derive p_3 and p_4, which means she cannot respond correctly to the question.*

[3]The questions "What is the measure of angle x?" and "Is angle x 30 degrees?" are different. The former suggests no specific value; its declarative form is not a proposition. In contrast, the declarative sentence corresponding to the latter, "Angle x is 30 degrees.", is a proposition. In this running example, agents are required to respond to the question about whether this proposition is true or false simply with "yes" or "no."

[4]f_1, f_2, and f_3 are theorems of a geometric theory but not logical truths. In this paper, we do not assume a theory such as Euclidean geometry; therefore, those mathematical facts are not valid formulae.

For representing each agent's awareness and knowledge in the example, we denote by E, I, and A the operators for explicit knowledge, implicit knowledge, and awareness, respectively. For a proposition φ, write $E_i\varphi$ for "agent i knows φ explicitly." A subscript of operators is omitted when the confusion does not occur, and E_i, for instance, is denoted simply by E. Following the definitions in FH logic, we define $E_i\varphi$ as $I_i\varphi \wedge A_i\varphi$ and $I_i\varphi$ as $K_i\varphi$ in EL.

Note that, although many propositions appear in this example, we pick up the last inference ("‡") that derives p_4, describing how implicit knowledge, explicit knowledge, and awareness are associated with the relevant propositions. Furthermore, let us emphasize that the focus of this discussion is on the agents' knowledge and awareness after their reasoning process is complete. We do not address dynamics such as becoming aware of new propositions through reasoning.

Agent a, who can construct the proof in Figure 2 in her mind, is aware of all the propositions; b, who stops the inference at "†", is unaware of p_3. Accordingly, each agent's awareness is expressed as follows:

(1) a is aware of p_2, p_3, f_3, p_4 ($A_a p_2, A_a p_3, A_a f_3, A_a p_4$);

(2) b is aware of p_2, f_3, p_4 ($A_b p_2, \neg A_b p_3, A_b f_3, A_b p_4$).

As for implicit knowledge, the agents have the following knowledge of the basics of elementary geometry from our assumption:

(3) a and b know f_3 implicitly ($I_a f_3$, $I_b f_3$);

(4) a and b implicitly know that if p_2, p_3, and f_3 hold, then p_4 holds, too ($I_a((p_2 \wedge p_3 \wedge f_3) \to p_4)$, $I_b((p_2 \wedge p_3 \wedge f_3) \to p_4)$).

Since both agents see the diagram and recognize each side and angle, each k_i belongs to their implicit knowledge. Hence, they also implicitly know p_2 and p_3. For it relies, eventually, solely on three assumptions: (i) Each f_i and k_i belongs to their implicit knowledge; (ii) the implication involved at each inference step itself belongs to their implicit knowledge; (iii) implicit knowledge is closed under implication.

(5) a and b implicitly know both p_2 and p_3 ($I_a p_2, I_a p_3, I_b p_2, I_b p_3$).

Furthermore, since implicit knowledge is closed under implication,

(6) a and b know p_4 implicitly ($I_a p_4$, $I_b p_4$).

(1), (2), and (6) derive $E_a p_4$ and $E_b p_4$ from the definition of explicit knowledge.

(7) a knows p_4 explicitly ($E_a p_4$).

(8) b knows p_4 explicitly ($E_b p_4$).

While the explicit knowledge (7) for a is consistent with the actual one, the knowledge (8) for b contradicts her available knowledge. The implicit knowledge of p_4 is justified through an inference from p_2, p_3, and f_3. Since b is unaware of p_3, this justification is not available within her awareness. Agent b thereby excludes p_4 from her explicit (available) knowledge. In other words, b, who is unaware of the required proposition p_3 to solve the problem, does not explicitly know that p_4 is true ($\neg E_b p_4$).

This resulting tension between the definition and intuition relates to the valid formula in FH logic, $I(\varphi) \wedge I(\varphi \rightarrow \psi) \rightarrow (A\psi \rightarrow E\psi)$. The formula is incoherent with the view that unless an agent is aware of all the propositions necessary to infer the consequence, it cannot be considered that she knows the consequence explicitly. For instance, for a system, we may take the deductive closure of its knowledge base as implicit knowledge and the information currently in the system's focus as awareness. Its explicit knowledge should be what is obtained through inference using propositions within the current focus, rather than what belongs merely to the deductively closed knowledge set currently focused on. This observation suggests that, in some cases, explicit knowledge cannot be captured merely by filtering unaware propositions out of implicit knowledge.

2.2 Awareness and Indistinguishability

The semantics of EL is presented by a Kripke model which consists of a non-empty set of possible worlds where each world has a different valuation for propositions, to which each agent may or may not be accessible. The knowledge is defined as a true proposition in all of an agent's accessible worlds. This accessibility is understood to express epistemic indistinguishability among possible worlds. For instance, an agent does not know that φ is true ($\neg K_i \varphi$) if and only if φ is true not in all her epistemically indistinguishable worlds (epistemic alternatives), which means that she cannot distinguish between the world where φ is true and the world where not.

The notion of awareness induces its own form of indistinguishability among possible worlds, as demonstrated in the following figure. Let p and q be propositions. Figure 3 illustrates a possible world model where circles express possible worlds, propositions listed above each world true propositions in the world, and a proposition within curly brackets an agent's aware proposition. The only difference between w and v is the truth value of q, and the agent is unaware of q in both worlds. In this situation, she views w and v as the same and does not distinguish between them due to a lack of awareness, which is represented by the dotted line. We call this

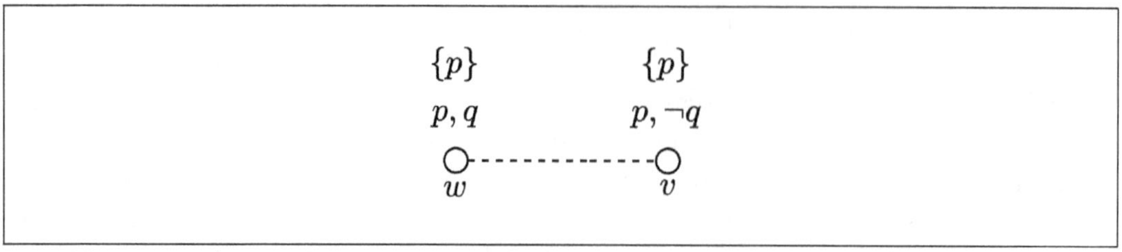

Figure 3: A possible world model where a set of possible worlds is $\{w, v\}$.

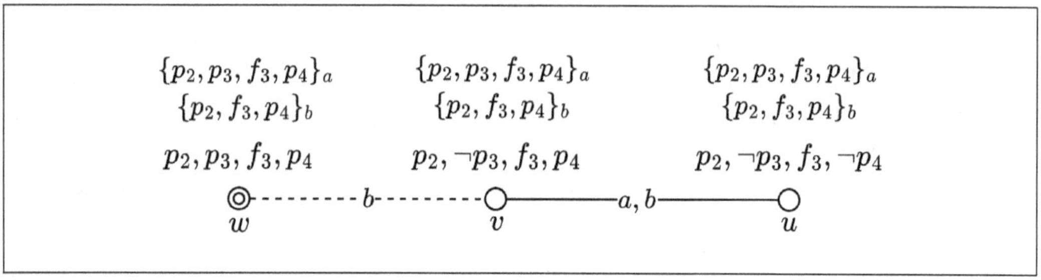

Figure 4: A possible world model where a set of possible worlds is $\{w, v, u\}$.

kind of indistinguishability specifically *awareness-indistinguishability* and show that it plays an important role in the formalization of explicit knowledge.

We consider which possible worlds for each agent are epistemically indistinguishable in the motivating example. Figure 4 illustrates a possible world model for Example 4, depicting three possible worlds and indistinguishability among them. For readability, we only include the four propositions p_2, p_3, f_3, and p_4 involved in the last inference ("‡"). The actual world is w, indicated by the double circle. Each curly bracket with a subscript of an agent indicates the agent's aware propositions. Since b is unaware only of p_3, she does not distinguish between w and v. This corresponds to awareness-indistinguishability, attributed to her lack of awareness, represented by the dotted line labeled with b in the figure. Furthermore, in v (where p_3 is false), neither agent knows p_4 even implicitly because of the absence of "‡", being unable to distinguish the world from u (where p_4 is false). This kind of indistinguishability, represented by the solid line labeled with a and b in the figure, expresses epistemic alternatives independent of the lack of awareness (in fact, these are still indistinguishable even under a's complete awareness). Note that since each agent has no way of knowing the other's awareness, we may consider the worlds differing only in the other's awareness from each of w, v, and u. However, we omit these, as, considering the presently discussed knowledge, they do not affect the description of agents' epistemic situations.

In the figure, the two kinds of indistinguishability jointly express that b cannot distinguish which world is the actual world among worlds, w, v (both in which $x = 30°$), and u (in which not), and, therefore, epistemic indistinguishability of b with incomplete awareness. The truth of knowledge is evaluated over all epistemic alternatives from the agent's perspective. The observation of Example 4 suggests that the epistemic alternatives for explicit knowledge should be presented through epistemic indistinguishability incorporating awareness-indistinguishability. Therefore, revisiting the conjunctive account of explicit knowledge ($E\varphi \leftrightarrow I\varphi \wedge A\varphi$), in the next section, we provide a formal definition of explicit knowledge through the composed epistemic indistinguishability and show that the formalization refines that in FH logic.

3 Awareness-Based Indistinguishability Logic

In this section, we formally define the syntax and semantics of our language, introducing our logic named *Awareness-Based Indistinguishability Logic* ($\mathcal{AIL}$). Note that when written in calligraphic letters, $\mathcal{AIL}$ refers to the set of formulae that are valid at every model. After showing the properties of the operators, we demonstrate that the formalization faithfully captures the knowledge in the example introduced in the previous section.

3.1 Language

Definition 1. *Let $\mathcal{P}$ be a countable set of atomic propositions and $\mathcal{G}$ a finite set of agents. The language $\mathcal{L}_{AIL}(\mathcal{P}, \mathcal{G})$ is the set of formulae generated by the following grammar:*

$$\mathcal{L}_{AIL}(\mathcal{P}, \mathcal{G}) \ni \varphi ::= p \mid \neg\varphi \mid \varphi \wedge \varphi \mid A_i\varphi \mid I_i\varphi \mid E_i\varphi \mid [\approx]_i\varphi \mid [\circ^+]_i\varphi,$$

where $p \in \mathcal{P}$ and $i \in \mathcal{G}$. Other logical connectives $\vee$, $\rightarrow$, and $\leftrightarrow$ are defined in the usual manner.

The operators A, I, and E represent awareness, implicit knowledge, and explicit knowledge operators, respectively.

- $A_i\varphi$ reads "agent i is aware of φ."
- $I_i\varphi$ reads "agent i knows φ implicitly."
- $E_i\varphi$ reads "agent i knows φ explicitly."

The other operators $[\approx]_i$ and $[\circ^+]_i$ are technical background ones. These are used to define explicit knowledge and play an important role in the proof of the completeness theorem in Section 5. Their detailed interpretations are explained in the following subsection.

3.2 Semantics

The semantics is presented in a Kripke style, primarily influenced by an awareness structure [7] and the equivalence relation proposed in [20].

Definition 2. *An epistemic model with awareness M is a tuple $\langle W, \{\sim_i, \mathscr{A}_i\}_{i \in \mathcal{G}}, V\rangle$, where:*

- *W is a non-empty set of possible worlds;*
- *$\sim_i$ is an equivalence relation on W;*
- *$\mathscr{A}_i : W \to 2^{\mathcal{P}}$ is an awareness function satisfying that if $(w, v) \in \sim_i$, then $\mathscr{A}_i(w) = \mathscr{A}_i(v)$;*
- *$V : \mathcal{P} \to 2^W$ is a valuation.*

This model consists of a Kripke model in $\mathcal{S}5$ and awareness functions[5]. In EL, the equivalence relations are simply called accessibility relations and define implicit knowledge. In our logic, we call it *IK-accessibility relation* to avoid confusion because some types of 'accessibilities' appear for defining explicit knowledge. The awareness function maps a world into a subset of atomic propositions which an agent is aware of at the world, and this subset $\mathcal{A}_i(w)$ is called an *awareness set* for the agent i. The restriction that awareness sets are equal across accessible worlds on $\sim_i$ ensures that agent i always knows her own awareness (awareness introspection), that is, $A_i\varphi \to I_i A_i \varphi$. We refer to this restriction as **ka** (knowledge implies awareness), following [22].

Moreover, we define another binary relation on W. The relation is associated with each agent's awareness set, which is specified by an awareness function, and, therefore, is model-dependent.

[5]This paper adopts $\mathcal{S}5$ (defining accessibility as an equivalence relation) as the foundation of models, for its tractability and widespread use in applications [8]. At the same time, $\mathcal{S}5$ has faced criticism for its introspection for knowledge [31]. However, the primary focus of this paper, as well as that of awareness logic in general, is to tackle the problem of logical omniscience, rather than the idealization of information processing associated with introspection. Using $\mathcal{S}5$ as the foundation allows us to isolate awareness from errors in information processing induced by lack of introspection, enabling a transparent analysis that highlights the role of awareness. Further details for introspection appear in Subsection 3.4.

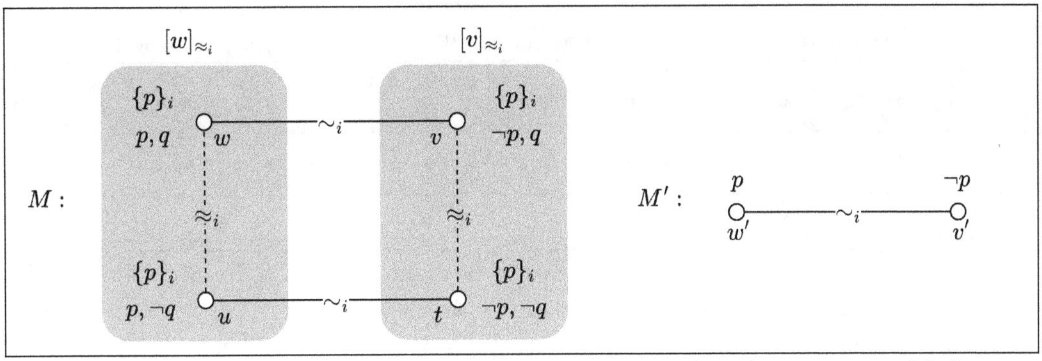

Figure 5: Two possible world models. Epistemic model with awareness M depicts equivalence classes by $\approx_i$. Agent i views w and u, and v and t as having no difference due to her lack of awareness of q. A Kripke model M' is the model only for i's aware proposition p. For readability, we omit the reflexivity.

Definition 3. *For each $i \in \mathcal{G}$, A-equivalence relation $\approx_i$ on W is defined by $(w, v) \in \approx_i$ iff, $\mathscr{A}_i(w) = \mathscr{A}_i(v)$ and $w \in V(p)$ iff $v \in V(p)$ for every $p \in \mathscr{A}_i(w)$.*

This relation groups possible worlds with different valuations only for unaware propositions and expresses awareness-indistinguishability. The relation is also an equivalence relation, and the equivalence classes collect worlds regarded as 'the same' from the agent's viewpoint at a given moment (Figure 5), in other words, as the same states for her current considering propositions. In model M in the figure, since proposition q is not in i's vocabulary, the subjective description of the agent's epistemic situation should be M' in which each equivalence class in M is viewed and 'simplified' as one possible world. We call an equivalence class by $\approx_i$, which is an element of the domain of the quotient model of M by $\approx_i$, a *simplified world for i*. Denoting an equivalence class of w by relation R_i by $[w]_{R_i}$, we can express a simplified world with respect to w for i as $[w]_{\approx_i}$. The restriction that awareness sets are equal across reachable worlds on $\approx_i$ ensures a property about i's knowledge of her own awareness, as well as that on $\sim_i$ does.

We express epistemic indistinguishability for explicit knowledge by composing IK-accessibility (indistinguishability independent of awareness) with awareness-indistinguishability (indistinguishability dependent on awareness). This resulting relation is called *EK-accessibility relation* and is achieved by taking the transitive closure of the composition of the two relations, $\sim_i$ and $\approx_i$. Formally, the relation is defined as $(\sim_i \circ \approx_i)^+$, where $\sim_i \circ \approx_i$ the sequential composition of $\approx_i$ and $\sim_i$, for a binary relation R, R^+ denotes the transitive closure of R, and the closure R^+ is the smallest set such that $R \subseteq R^+$ and for all x, y, z, if $(x, y) \in R^+$ and $(y, z) \in R^+$,

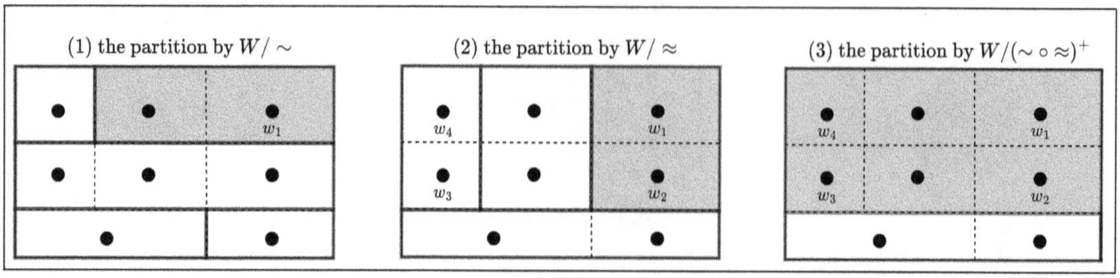

Figure 6: Partitions of a set W consisting of eight worlds by IK-accessibility, A-equivalence, and EK-accessibility relations. Given $[w_1]_{\sim_i}$ and $[w_1]_{\approx_i}$ (see (1) and (2)), w_2 in $[w_1]_{\approx_i}$ belongs to the partition by $(\sim_i \circ \approx_i)^+$. Then, w_3 is in this connected partition as $w_3 \sim_i w_2$. Transitively, w_4 is also in it as $w_3 \approx_i w_4$. The final partition becomes as (3).

then $(x, z) \in R^+$. As shown in Lemma 1, this closure is also an equivalence relation on W and partitions W more coarsely (Figure 6).

Lemma 1. *Let $\sim_i$ be IK-accessibility relation and $\approx_i$ A-equivalence relation. EK-accessibility relation $(\sim_i \circ \approx_i)^+$ is an equivalence relation.*

Proof. For the reflexivity and the transitivity, the proof is straightforward because both $\sim_i$ and $\approx_i$ are equivalence relations and $(\sim_i \circ \approx_i)^+$ is closed by transitivity. For the symmetricity, It is sufficient to prove, for every $w, v \in W$, if $(w, v) \in (\sim_i \circ \approx_i)^+$, then $(v, w) \in (\sim_i \circ \approx_i)^+$. It follows from the assumption that there exists $u_1, \ldots, u_n \in W$ such that $(w, u_1) \in \sim_i \circ \approx_i$, $(u_1, u_2) \in \sim_i \circ \approx_i$, $\cdots$, $(u_{n-1}, u_n) \in \sim_i \circ \approx_i$, and $(u_n, v) \in \sim_i \circ \approx_i$. In addtion, for each pair, we take (w, u_1) as an example, there exists $t \in W$ such that $(w, t) \in \approx_i$ and $(t, u_1) \in \sim_i$. Since both relations are equivalence relations, $(u_1, u_1), (t, w) \in \approx_i$, $(u_1, t), (w, w) \in \sim_i$, and $(u_1, t), (t, w) \in \sim_i \circ \approx_i$. Thus, there exists $t_i \in W$, where $1 \leq i \leq n+1$, such that $(u_1, t_1), (t_1, w) \in \sim_i \circ \approx_i$, $(u_2, t_2), (t_2, u_1) \in \sim_i \circ \approx_i$, $\cdots$, $(u_n, t_n), (t_n, u_{n-1}) \in \sim_i \circ \approx_i$, and $(v, t_{n+1}), (t_{n+1}, u_n) \in \sim_i \circ \approx_i$. Therefore, by the transitivity, $(v, w) \in (\sim_i \circ \approx_i)^+$. □

However, in mathematical authenticity, these two relations, $\sim_i$ and $\approx_i$, belong to different concepts. In order to avoid a misleading formalization, we may regard that (i) $\approx_i$ is an ϵ-closure of automata, where we can freely move to different states (worlds) without any computational procedure [23]. Otherwise, (ii) we may regard that the set of possible worlds W is re-modelled to its quotient space $W/\approx_i$ and after that we may introduce the equivalence $\sim_i$. Nevertheless, because awareness-indistinguishability is static and does not involve any update or re-modelling in epistemic states of agents, we adopt the present formalization.

The satisfaction relation is defined by a pointed model, a pair (M, w) of a model and a world. Let $At(\varphi)$ denote the set of atomic propositions occurring in φ.

Definition 4. *For each epistemic model with awareness M and possible world $w \in W$, the satisfaction relation $\models_{AIL}$ is given as follows:*

$$
\begin{array}{rl}
M, w \models_{AIL} p & \textit{iff } w \in V(p); \\
M, w \models_{AIL} \neg\varphi & \textit{iff } M, w \not\models_{AIL} \varphi; \\
M, w \models_{AIL} \varphi \wedge \psi & \textit{iff } M, w \models_{AIL} \varphi, \textit{ and } M, w \models_{AIL} \psi; \\
M, w \models_{AIL} A_i\varphi & \textit{iff } At(\varphi) \subseteq \mathscr{A}_i(w); \\
M, w \models_{AIL} I_i\varphi & \textit{iff } M, v \models_{AIL} \varphi \textit{ for all } v \textit{ such that } (w, v) \in \sim_i; \\
M, w \models_{AIL} [\approx]_i\varphi & \textit{iff } M, v \models_{AIL} \varphi \textit{ for all } v \textit{ such that } (w, v) \in \approx_i; \\
M, w \models_{AIL} [\circ^+]_i\varphi & \textit{iff } M, v \models_{AIL} \varphi \textit{ for all } v \textit{ such that } (w, v) \in (\sim_i \circ \approx_i)^+; \\
M, w \models_{AIL} E_i\varphi & \textit{iff } M, w \models_{AIL} A_i\varphi \textit{ and } M, w \models_{AIL} [\circ^+]_i\varphi.
\end{array}
$$

When the confusion does not occur, $\models_{AIL}$ is denoted simply by $\models$.

Each modal operator corresponds to a distinct structural component of the model. In what follows, we summarize their formal definitions and provide intuitive explanations.

- By the definition, the awareness of φ ($A_i\varphi$) is reduced to awareness for the atomic propositions occurring in φ and then interpreted. For instance, "an agent is aware of $\varphi \wedge \psi$" if and only if "she is aware of φ and ψ." This interpretation with the restriction called **gpp** (awareness generated by primitive propositions) [22] provides a representation of awareness in the sense of *awareness of*.

- The $[\approx]_i$ operator refers to a true proposition in all i's awareness-indistinguishable worlds or a simplified world for i. $[\approx]_i\varphi$ is true at world w if and only if φ is true in all worlds in which each aware proposition has the same truth value as in w, unaware propositions disregarded. A formula with this modality added can be understood as the formula with the modality removed in a simplified world. For, $[\approx]_i\varphi$ holds at w precisely when φ is true in the simplified world with respect to w for i. In other words, this operator refers to equivalence classes representing an 'abstraction' of the given model [24]. A deeper analysis of the operator lies beyond our present scope and is left for future work.

- The $[\circ^+]_i$ operator refers to a true proposition in all i's indistinguishable worlds in terms of those two kinds of indistinguishability, that is, on the EK-accessibility relation. Using the composition instead of the union makes $[\circ^+]_i$ be equivalent to infinite iteration of $I_i[\approx]_i$. Although a formula with infinite

	FH Logic[6]	Our Proposal
Awareness	$A_i\varphi :\Longleftrightarrow At(\varphi) \in \mathscr{A}_i(w)$	
Implicit Knowledge	$I_i\varphi :\Longleftrightarrow w \sim_i v$	
A-equivalence		$[\approx]_i\varphi :\Longleftrightarrow w \approx_i v$
EK-accessibility		$[\circ^+]_i\varphi :\Longleftrightarrow w(\sim_i \circ \approx_i)^+ v$
Explicit Knowledge	$E_i\varphi :\Longleftrightarrow A_i\varphi \wedge I_i\varphi$	$E_i\varphi :\Longleftrightarrow A_i\varphi \wedge [\circ^+]_i\varphi$

Table 1: The outline of the differences in formalizations between an FH logic and our logic. The double-lined arrow $:\Longleftrightarrow$ with a colon indicates "is defined by." Note that for readability, some details of the conditions, such as a model and world, are omitted.

iteration is not well-formed, this observation serves to prove the completeness theorem.

- Implicit knowledge is interpreted similarly to K of EL, through IK-accessibility relation. On the other hand, explicit knowledge is interpreted by an agent's awareness and her epistemically indistinguishable worlds referred to by $[\circ^+]_i$ operator, not by I_i operator.

Our logic does not validate $I_i\varphi \wedge A_i\varphi \rightarrow E_i\varphi$, unlike FH logic. For example, consider model M in Figure 5. Propositions A_ip, I_ip, and $[\approx]_ip$ hold at w but not $[\circ^+]_ip$. Therefore, E_ip does not hold at w. Table 1 outlines the differences in its formalizations. We will discuss some properties of our operators in Subsection 3.4.

The validity is defined as usual.

Definition 5. *A formula φ is valid at M if φ holds at every pointed world in M, which is denoted by $M \vDash \varphi$. A formula φ is valid if φ holds at every pointed model, which is denoted by $\vDash \varphi$.*

3.3 An Explanation with Dynamic Operators

We discuss more deeply what the information represented by $[\circ^+]$ refers to. The definition clearly indicates that this is stronger than implicit knowledge. However, when an agent is aware of all propositions, both are equivalent to explicit knowledge. To demonstrate the difference between them, we temporarily introduce two simple

[6] A FH logic with the restriction known as **gpp** [22]. This restriction reduces awareness of a formula to that of the atomic propositions occurring therein and corresponds to the notion of awareness that this paper focuses on, as mentioned in Section 1. Further details appear in Section 4.

dynamic operators: a becoming-aware operator and a becoming-unaware operator. These operators have no other functionality besides adding propositions to, or removing them from, an agent's awareness set. Formally, let $\mathcal{Q} \subseteq \mathcal{P}$, and the updated models $M[+\mathcal{Q}]_i$ and $M[-\mathcal{Q}]_i$ are defined as follows: $M[+\mathcal{Q}]_i$ is an epistemic modal with awareness $\langle W, \{\sim_i, \mathscr{A}'_i\}_{i\in\mathcal{G}}, V\rangle$, where,

- for all w, $\mathscr{A}'_i(w) := \mathscr{A}_i(w) \cup \mathcal{Q}$;

$M[-\mathcal{Q}]_i$ is an epistemic modal with awareness $\langle W, \{\sim_i, \mathscr{A}'_i\}_{i\in\mathcal{G}}, V\rangle$, where,

- for all w, $\mathscr{A}'_i(w) := \mathscr{A}_i(w) \setminus \mathcal{Q}$.

The satisfaction relation for each operator is as follows:

$$M, w \vDash [+\mathcal{Q}]_i\varphi \textit{ iff } M[+\mathcal{Q}]_i, w \vDash \varphi;$$
$$M, w \vDash [-\mathcal{Q}]_i\varphi \textit{ iff } M[-\mathcal{Q}]_i, w \vDash \varphi.$$

Using these operators, we find that $[\circ^+]\varphi$ implies $[+At(\varphi)]E[-At(\varphi)]\varphi$ ($[\circ^+]\varphi \to [+At(\varphi)]E[-At(\varphi)]\varphi$). That is, the information referred to by $[\circ^+]$ would become explicit knowledge immediately upon becoming aware of that particular proposition. On the other hand, implicit knowledge requires awareness of all propositions. Formally, $I\varphi \leftrightarrow [+\mathcal{P}]E([-\mathcal{P}]\varphi)$. For the implicit knowledge to become explicit, awareness of that proposition alone is insufficient; awareness of all propositions is required. These formulae highlight the distinctions between I and $[\circ^+]$ operators.

Furthermore, this observation is coherent with the interpretation of implicit knowledge as the ideal knowledge of a logically omniscient agent and corroborates our definition of explicit knowledge. For, the possession of knowledge encompassing all tautologies, along with knowledge derived from the agent's existing knowledge through logical inference, that is, implicit knowledge, is appropriate for an agent capable of referring to, being fully aware of, all propositions at a given time. Such an agent can evaluate the necessary truth of any proposition, including tautologies, and perform logical inference on any given propositions.

3.4 Basic Properties

In this subsection, we show some properties of the operators introduced in the previous subsection, mainly A, I, and E, including their interactions.

Awareness The following validity is based on the definition of the awareness operator.

- $\vdash A_i\varphi \leftrightarrow \bigwedge_{p\in At(\varphi)} A_ip$.

The formula means that being aware of a proposition φ is equivalent to being aware of the atomic propositions occurring in φ, capturing the notion of *awareness of* as mentioned in Section 1.

The definitions of $\sim_i$ and $\approx_i$ have the conditions on the equality of awareness sets, from which the following expressions hold:

- $\models I_i A_i \varphi$ and $\models I_i \neg A_i \varphi$,
- $\models E_i A_i \varphi$ and $\models \neg E_i \neg A_i \varphi$.

These validities show that awareness is always implicitly and explicitly known, while unawareness does not belong to the agent's explicit knowledge.

Explicit and Implicit knowledge Explicit knowledge in FH logic is defined as the aware implicit knowledge. On the other hand, the following expressions hold in our model:

- $\models E_i\varphi \to I_i\varphi \wedge A_i\varphi$,
- $\not\models I_i\varphi \wedge A_i\varphi \to E_i\varphi$,
- For all $w \in W$ in M, if $\mathcal{P} = \mathscr{A}_i(w)$, then $M \models I_i\varphi \leftrightarrow E_i\varphi$,

In our logic, explicit knowledge implies implicit knowledge and awareness, but the converse does not hold. This invalidity (the second formula) indicates the case, as shown in Example 4, where an agent knows a proposition implicitly and is aware of that proposition itself, yet it is still not explicitly known. This is one of the apparent differences between FH and our logic, showing that our explicit knowledge is a refinement of that in FH logic. Nevertheless, the third expression holds, as with FH logic. This means that if an agent is aware of all propositions, implicit knowledge is equivalent to explicit knowledge.

Accordingly, we find the following formula, which is valid in FH logic, to be invalid.

- $\not\models I_i\varphi \wedge I_i(\varphi \to \psi) \to (A_i\psi \to E_i\psi)$.

This formula itself means that even when an agent lacks awareness of an inference process, as long as she is aware of its consequence, she classifies that consequence as explicit knowledge. Our logic excludes this formula. In the view introduced in Section 1, the agent does not know the consequence explicitly unless she is aware of all the propositions necessary to infer the consequence.

Introspection Both implicit and explicit knowledge in our logic are interpreted with equivalence relations; hence, for I and E, the axioms T and 4 hold.

- $\models I_i\varphi \rightarrow \varphi$ and $\models E_i\varphi \rightarrow \varphi$,
- $\models I_i\varphi \rightarrow I_iI_i\varphi$ and $\models E_i\varphi \rightarrow E_iE_i\varphi$ (Positive introspection).

However, the axiom 5 does not hold for the explicit knowledge. The axiom represents negative introspection: an agent knowing what she does not know. This introspection is viewed as unrealistic for human knowledge.

- $\models \neg I_i\varphi \rightarrow I_i\neg I_i\varphi$ (Negative introspection),
- $\not\models \neg E_i\varphi \rightarrow E_i\neg E_i\varphi$,
- $\models \neg E_i\varphi \wedge A_i\varphi \rightarrow E_i\neg E_i\varphi$ (Weak negative introspection).

The third formula, known as weak negative introspection, is valid, stating that an agent explicitly knows what she does not know explicitly only when she is aware of that proposition. Our logic is $\mathcal{S}5$-based because of several benefits (See the footnote in Definition 2); therefore, as seen, the logic has the property of introspection. This shows that while our awareness logic does address one aspect of the concern that $\mathcal{S}5$ is 'too strong' for human knowledge by avoiding logical omniscience, which has been our primary focus, it scarcely addresses the aspect concerning introspection [31].

From the definitions, $[\circ^+]_i\varphi$ is equivalent to the infinite formula $(I_i[\approx]_i)^1\varphi \wedge \dots$, where $(I_i[\approx]_i)^n$ is iteration of $I_i[\approx]_i$ n times, therefore $E_i\varphi$ is equivalent to the infinite formula $A_i\varphi \wedge (I_i[\approx]_i)^1\varphi \wedge \dots$. Although these formulae are not well-formed, the following expressions follow:

- $\not\models I_i\varphi \rightarrow E_iI_i\varphi$ (Mix4),
- $\not\models \neg I_i\varphi \rightarrow E_i\neg I_i\varphi$ (Mix5),
- $\not\models I_i\varphi \wedge A_i\varphi \rightarrow E_iI_i\varphi$ (Mix4A),
- $\models \neg I_i\varphi \wedge A_i\varphi \rightarrow E_i\neg I_i\varphi$ (Mix5A).

The first and second formulae express that explicit knowledge of (or about) implicit knowledge/ignorance cannot be derived from the implicit information alone. The third formula, which prefixes awareness to the antecedent of the axiom 4, is invalid and shows that implicit knowledge and awareness are insufficient for obtaining explicit knowledge again. On the other hand, the fourth formula, which prefixes awareness to the antecedent of 5, is valid.

Logical omniscience As with other logics of awareness, our explicit knowledge avoids characterizations of logical omniscience.

- $\models \varphi$ does not imply $\models E_i\varphi$ (Knowledge of valid formulae),
- $M, w \models E_i\varphi$ and $\models \varphi \rightarrow \psi$ do not imply $M, w \models E_i\psi$ (Closure under logical implication).
- $M, w \models E_i\varphi$ and $\models \varphi \leftrightarrow \psi$ do not imply $M, w \models E_i\psi$ (Closure under logical equivalence).

Furthermore, the expressions with awareness prefixed to the antecedent hold.

- $\models \varphi$ and $M, w \models A_i\varphi$ imply $M, w \models E_i\varphi$,
- $M, w \models E_i\varphi \wedge A_i\psi$ and $\models \varphi \rightarrow \psi$ imply $M, w \models E_i\psi$
- $M, w \models E_i\varphi \wedge A_i\psi$ and $\models \varphi \leftrightarrow \psi$ imply $M, w \models E_i\psi$

Implication relations Some operators imply other operators. The following expressions show some implication relations among modalities.

- $\models E_i\varphi \leftrightarrow A_i\varphi \wedge [\circ^+]_i\varphi$,
- $\models [\circ^+]_i\varphi \rightarrow [\approx]_i I_i\varphi \wedge I_i\varphi \wedge [\approx]_i\varphi$,
- $\models I_i\varphi \wedge A_i\varphi \rightarrow [\approx]_i\varphi$,
- $\not\models [\approx]_i\varphi \rightarrow I_i\varphi$ and $\not\models [\approx]_i\varphi \rightarrow A_i\varphi$.

The $[\circ^+]$ operator is implied by the E operator and implies not only the I and $[\approx]$ operators but also $[\approx]I$. Thus, it is a necessary condition for φ to be included in explicit knowledge that φ remains in the knowledge of i as a logically omniscient agent ($I_i\varphi$), even in the simplified world in which propositions of which i is unaware are disregarded. The $[\approx]$ operator is implied by the I and A operators, but the converse is not. Arranging the operators in the order, we have

- $E = [\circ^+] + A > [\approx]I + A > I + A > [\approx]$,

where $X = Y$ abbreviates $X\varphi \leftrightarrow Y\varphi$ and $X > Y$ $X\varphi \rightarrow Y\varphi$ for all φ.

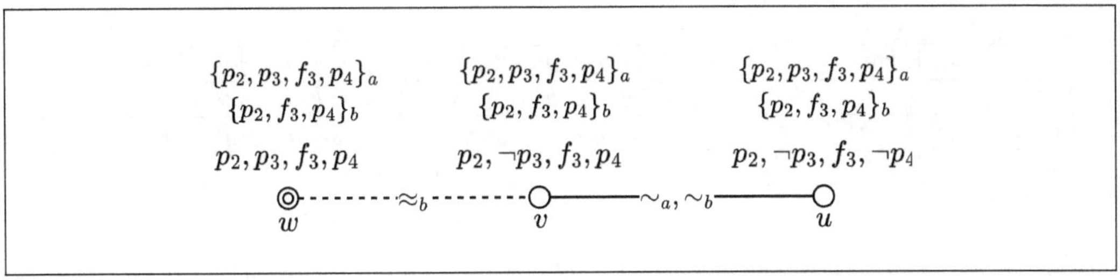

Figure 7: An epistemic model with awareness for Example 4. For readability, we omit the reflexivity.

3.5 A Representation of the Motivating Example

We provide a formal representation of the agents' knowledge in Example 4 with our semantics.

We recall Example 4 in Section 2. The last inference in Figure 2 involves p_2, p_3, f_3, and p_4, each of which reads:

- p_2: "The lengths of side DA and DC are equal;"
- p_3: "The sum of the measures of angle DAC and DCA is 60 degrees;"
- f_3: "The measures of angles opposite the equal sides are equal;"
- p_4: "The measure of angle x is 30 degrees."

Reflecting their awareness and knowledge, the IK-accessibility relations should be:

- $\sim_a = \sim_b = \{(w, w), (v, v), (u, u), (v, u), (u, v)\}$.

As for awareness, since b is unaware of p_2 at all worlds,

- $\mathscr{A}_a(w) = \mathscr{A}_a(v) = \mathscr{A}_a(u) = \{p_2, p_3, f_3, p_4\}$;
- $\mathscr{A}_b(w) = \mathscr{A}_b(v) = \mathscr{A}_b(u) = \{p_2, f_3, p_4\}$;

From the definition, the A-equivalence relations should be:

- $\approx_a = \{(w, w), (v, v), (u, u)\}$;
- $\approx_b = \{(w, w), (v, v), (u, u), (w, v), (v, w)\}$;

	Agent a	Agent b
Implicit knowledge	f_i,k_i,p_i,$p_2 \wedge p_3 \wedge f_3 \rightarrow p_4$	f_i,k_i,p_i,$p_2 \wedge p_3 \wedge f_3 \rightarrow p_4$
Awareness	f_i,k_i,p_i,$p_2 \wedge p_3 \wedge f_3 \rightarrow p_4$	f_1,f_3,k_i,p_1,p_2,p_4
EK-accessibility ($[\circ^+]$)	f_i,k_i,p_i,$p_2 \wedge p_3 \wedge f_3 \rightarrow p_4$	f_1,f_3,k_i,p_1,p_2,$p_2 \wedge p_3 \wedge f_3 \rightarrow p_4$
Explicit knowledge in FH logic	f_i,k_i,p_i,$p_2 \wedge p_3 \wedge f_3 \rightarrow p_4$	f_1,f_3,k_i,p_1,p_2,p_4
Our explicit knowledge	f_i,k_i,p_i,$p_2 \wedge p_3 \wedge f_3 \rightarrow p_4$	f_1,f_3,k_i,p_1,p_2

Table 2: Implicit knowledge, explicit knowledge, and awareness of each agent in Example 4. Note that this table is not exhaustive; for instance, a logical implication involved in "†" and those such as $k_1 \wedge k_2$ are not included.

These are illustrated in Figure 7. In the actual world w, implicit knowledge, awareness, and explicit knowledge of each agent ((1),...,(7) and $\neg E_b p_4$ instead of (8)) are correctly represented, that is, for b, we have $M, w \models \neg A_b p_3 \wedge A_b(p_2 \wedge f_3 \wedge p_4) \wedge I_b(p_2 \wedge p_3 \wedge f_3) \wedge I_b(p_2 \wedge p_3 \wedge f_3 \rightarrow p_4) \wedge I_b p_4 \wedge \neg E_b p_4$. This is consistent with the intuition about b's explicit knowledge, which is that agent b, who is unaware of the required proposition to answer the question, does not explicitly know the answer. As for a, we have $M, w \models E_a p_4$, and the consistency with her knowledge is maintained.

4 A Comparison with FH logic: Expressivity and Embedding

This section compares our logic with an FH logic in terms of expressivity and then proves that the FH logic is embeddable into our logic.

FH logic has all the formulae of $\mathcal{L}_{AIL}(\mathcal{P}, \mathcal{G})$ excluding $[\approx]$ and $[\circ^+]$ operators. Formally, the language $\mathcal{L}_{FH}(\mathcal{P}, \mathcal{G})$ is the set of formulae generated by the following grammar:

$$\mathcal{L}_{FH}(\mathcal{P}, \mathcal{G}) \ni \varphi ::= p \mid \neg\varphi \mid \varphi \wedge \varphi \mid A_i\varphi \mid I_i\varphi \mid E_i\varphi,$$

where $p \in \mathcal{P}$ and $i \in \mathcal{G}$. Other logical connectives $\vee$, $\rightarrow$, and $\leftrightarrow$ are defined in the usual manner. When it is unambiguous from the context, we omit $\mathcal{P}, \mathcal{G}$ and write $\mathcal{L}_{AIL}$ instead of $\mathcal{L}_{AIL}(\mathcal{P}, \mathcal{G})$ and $\mathcal{L}_{FH}$ $\mathcal{L}_{FH}(\mathcal{P}, \mathcal{G})$. To avoid confusion, superscripts indicating the languages are attached to operators in formulae, such as E_i^{FH} and E_i^{AIL}.

FH logic is usually interpreted by awareness structures. Here, we focus on the FH logic interpreted by the structure with restrictions regarding awareness and knowledge, **gpp** and **ka**, which are focused on in most applications [22, 4]. Under the restrictions **gpp** and **ka**, the resulting awareness structure is structurally identical to an epistemic model with awareness. Therefore, the semantics can also be given by an epistemic model with awareness, as with $\mathcal{AIL}$. The satisfaction relation $\models_{FH}$ is defined in the same way as that in $\mathcal{AIL}$, except for the E operator.

$$M, w \models_{FH} E_i\varphi \textit{ iff } M, w \models_{FH} A_i\varphi \text{ and } M, w \models_{FH} I_i\varphi,$$

First, we prove that our logical language is at least as expressive as that of the FH logic.

Definition 6. *Two formulae φ and ψ are equivalent, if for every model and world, $M, w \models \varphi$ iff $M, w \models \psi$, which is denoted by $\varphi \equiv \psi$. A language $\mathcal{L}_2$ is at least as expressive as $\mathcal{L}_1$, if for every $\varphi_1 \in \mathcal{L}_1$, there is $\varphi_2 \in \mathcal{L}_2$ such that $\varphi_1 \equiv \varphi_2$, which is denoted by $\mathcal{L}_1 \preceq \mathcal{L}_2$. $\mathcal{L}_2$ is more expressive than $\mathcal{L}_1$, if $\mathcal{L}_1 \preceq \mathcal{L}_2$ and $\mathcal{L}_2 \not\preceq \mathcal{L}_1$, which is denoted by $\mathcal{L}_1 \prec \mathcal{L}_2$.*

Lemma 2. $\mathcal{L}_{FH} \preceq \mathcal{L}_{AIL}$.

Proof. It is sufficient to show a translation $t : \mathcal{L}_{FH} \to \mathcal{L}_{AIL}$ such that $t(\varphi)$ is equivalent to φ for every $\varphi \in \mathcal{L}_{FH}$. This is proved by induction on the structure of formulae. The only non-trivial case is the E operator. The translation for this operator is provided by $t(E_i^{FH}\varphi) = A_i^{AIL}t(\varphi) \wedge I_i^{AIL}t(\varphi)$. □

Thus, for every formula of the FH logic, there exists a formula of $\mathcal{AIL}$ that has the same meaning as that formula.

To prove having more expressive power, we first introduce bisimulation [30] between our models and show that this relation implies modal equivalence with respect to the FH logic. This bisimulation is a special case of *standard bisimulation* proposed by [29] and suffices for our purpose.

Definition 7. *A bisimulation between epistemic models with awareness $M = \langle W, \{\sim_i, A_i\}_{i\in\mathcal{G}}, V\rangle$ and $M' = \langle W', \{\sim'_i, A'_i\}_{i\in\mathcal{G}}, V'\rangle$ is a relation B on $W \times W'$ satisfying that, for every $p \in \mathcal{P}$ and $i \in \mathcal{G}$, if $(w, w') \in B$, then:*

- *$w \in V(p)$ iff $w' \in V'(p)$;*
- *if $(w, v) \in \sim_i$, then there exists v' such that $(w', v') \in \sim'_i$ and $(v, v') \in B$;*
- *if $(w', v') \in \sim'_i$, then there exists v such that $(w, v) \in \sim_i$ and $(v, v') \in B$;*
- *$\mathscr{A}_i(w) = \mathscr{A}'_i(w')$.*

We write $(M, w) \rightleftharpoons (M', w')$, if and only if there exists a bisimulation B between M and M' such that $(w, w') \in B$.

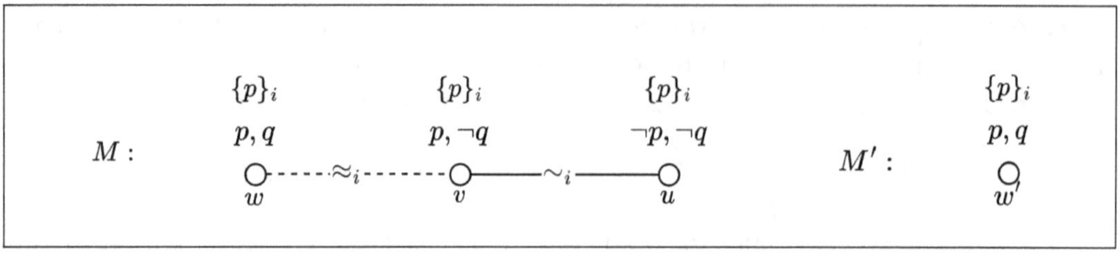

Figure 8: Two epistemic models with awareness for a single agent i. Pointed model (M, w) is bisimilar to (M', w'). For readability, we omit the reflexivity.

Lemma 3. *For all every $\varphi \in \mathcal{L}_{FH}$, if $(M, w) \rightleftharpoons (M', w')$, then $(M, w) \vDash_{FH} \varphi$ iff $(M', w') \vDash_{FH} \varphi$.*

Proof. We prove it by induction on the structure of formulae. As for the base case and that of logical connectives, these proofs are straightforward from the first condition of the bisimulation.

- For the case of $I_i\varphi$, suppose that $(M, w) \vDash_{FH} I_i\varphi$. For any v' such that $(w', v') \in \sim_i'$, from the third condition, there exists v such that $(w, v) \in \sim_i$ and $(v, v') \in B$. Given that for all v such that $(w, v) \in \sim_i$, $M, v \vDash_{FH} \varphi$, it follows from the induction hypothesis that $M', v' \vDash_{FH} \varphi$. Since v' is an arbitrary world such that $(w', v') \in \sim_i'$, thus, $M', w' \vDash_{FH} I_i\varphi$. The converse direction can be proved similarly by the second condition.
- For the case of $A_i\varphi$, suppose that $(M, w) \vDash_{FH} A_i\varphi$, then $At(\varphi) \subseteq \mathscr{A}_i(w)$. From the fourth condition, $\mathscr{A}_i(w) = \mathscr{A}_i'(w')$. Thus, $At(\varphi) \subseteq \mathscr{A}_i(w')$ and $(M', w') \vDash_{FH} A_i\varphi$.
- For the case of $E_i\varphi$, It can be proved easily by decomposing into $A_i\varphi$ and $I_i\varphi$.

□

Lemma 4. $\mathcal{L}_{AIL} \not\preceq \mathcal{L}_{FH}$.

Proof. Consider M and M' in Figure 8. Model M consists of three worlds and M' of a single world. Since $(M, w) \leftrightarrows (M', w')$, for every $\varphi \in \mathcal{L}_{FH}$, $M, w \vDash_{FH} \varphi$ iff $M', w' \vDash_{FH} \varphi$. On the other hand, $M, w \vDash_{AIL} E_ip$ and $M', w' \vDash_{AIL} \neg E_ip$. Therefore, no formula equivalent to E_ip exists. □

Theorem 1. $\mathcal{L}_{FH} \prec \mathcal{L}_{AIL}$.

Proof. By Lemma 2 and 4, trivial. □

Furthermore, we prove that the FH logic is embeddable into $\mathcal{AIL}$ by utilizing the translation constructed in Lemma 2.

Theorem 2. *For every* $\varphi \in \mathcal{L}_{FH}$, $\models_{FH} \varphi$ *iff* $\models_{AIL} t(\varphi)$.

Proof. We construct a translation $t : \mathcal{L}_{FH} \to \mathcal{L}_{AIL}$ such that for all models and worlds, $M, w \models_{FH} \varphi$ *iff* $M, w \models_{AIL} t(\varphi)$ as well as Lemma 2. This completes the proof. □

5 Hilbert System AIL

We now move on to the proof theory. The operators for awareness, implicit knowledge, and $[\approx]$ can be axiomatized based on the axiomatic system[7] **S5** [6] and that of FH logic [14]. Even though all of the operators apart from A are necessity operators on equivalence relations, the $[\circ^+]$ operator has interactions with other operators and makes the compactness of $\mathcal{AIL}$ lost; therefore, its complete axiomatization cannot be a mere **S5** with awareness and multimodal operators.

Table 3 presents a Hilbert system **AIL**. The seven axioms about awareness AN, AC, AA, AI, A$[\approx]$, A$[\circ^+]$M, and AE reduce awareness for complex formulae into awareness for atomic propositions forming that formula, corresponding to **gpp**. The axioms I$[\approx]$A and I$[\approx]$NA correspond to **ka**. The axiom $AA[\approx]$ reflects the definition of A-equivalence relation and means that an aware true proposition at world w remains in all worlds regarded as the same worlds as w from the agent's viewpoint. For $\mathrm{K_I}, \mathrm{T_I}, 5_\mathrm{I}, \mathrm{K}_{[\approx]}, \mathrm{T}_{[\approx]}$, and $5_{[\approx]}$, as I and $[\approx]$ are necessity operators on equivalence relations, we adopt K,T, and 5 in modal logic for these. The axioms for the $[\circ^+]$ operator, $\mathrm{K}_{[\circ^+]}$, MIX, and IND, are based on axioms of a logic with common knowledge [30] because both operators are interpreted by reflexive-transitive closures. Together, these lead to the validity of T and 4 for the E operator. The axiom EA$[\circ^+]$ corresponds to the requirement for the E operator on the satisfaction relation.

Definition 8. *A system* **AIL** *is a set of formulae containing the axioms in Table 3 and is closed under the inference rules in the table. We write* $\vdash \varphi$ *if* $\varphi \in$ **AIL**. *Let* Γ *be a set of formulae in* $\mathcal{L}_{AIL}$ *and* $\bigwedge \Gamma$ *be an abbreviation of* $\bigwedge_{\varphi \in \Gamma} \varphi$. *If there is a finite subset* Γ' *of* Γ *such that* $\vdash \bigwedge \Gamma' \to \varphi$, *we write* $\Gamma \vdash \varphi$ *and call it that* φ *is deducible from* Γ.

[7] When written in bold face, **S5** refers to the axiomatic system, distinguished from the class $S5$ of models.

Axiom schemata	
TAUT	The set of propositional tautologies
AN	$A_i\varphi \leftrightarrow A_i\neg\varphi$
AC	$A_i(\varphi \wedge \psi) \leftrightarrow A_i\varphi \wedge A_i\psi$
AA	$A_i\varphi \leftrightarrow A_iA_j\varphi$
AI	$A_i\varphi \leftrightarrow A_iI_j\varphi$
A[≈]	$A_i\varphi \leftrightarrow A_i[\approx]_j\varphi$
A[∘+]	$A_i\varphi \leftrightarrow A_i[\circ^+]_j\varphi$
AE	$A_i\varphi \leftrightarrow A_iE_j\varphi$
IA	$A_i\varphi \rightarrow I_iA_i\varphi$
INA	$\neg A_i\varphi \rightarrow I_i\neg A_i\varphi$
AA[≈]	$A_ip \wedge p \rightarrow [\approx]_ip$
K_I	$I_i(\varphi \rightarrow \psi) \rightarrow (I_i\varphi \rightarrow I_i\psi)$
T_I	$I_i\varphi \rightarrow \varphi$
5_I	$\neg I_i\varphi \rightarrow I_i\neg I_i\varphi$
$K_{[\approx]}$	$[\approx]_i(\varphi \rightarrow \psi) \rightarrow ([\approx]_i\varphi \rightarrow [\approx]_i\psi)$
$T_{[\approx]}$	$[\approx]_i\varphi \rightarrow \varphi$
$5_{[\approx]}$	$\neg[\approx]_i\varphi \rightarrow [\approx]_i\neg[\approx]_i\varphi$
$K_{[\circ^+]}$	$[\circ^+]_i(\varphi \rightarrow \psi) \rightarrow ([\circ^+]_i\varphi \rightarrow [\circ^+]_i\psi)$
MIX	$[\circ^+]_i\varphi \rightarrow \varphi \wedge [\approx]_iI_i[\circ^+]_i\varphi$
IND	$[\circ^+]_i(\varphi \rightarrow [\approx]_iI_i\varphi) \rightarrow (\varphi \rightarrow [\circ^+]_i\varphi)$
EA[∘+]	$E_i\varphi \leftrightarrow A_i\varphi \wedge [\circ^+]_i\varphi$
Inference Rules	
MP	If $\vdash \varphi$ and $\vdash \varphi \rightarrow \psi$, then $\vdash \psi$
GI	If $\vdash \varphi$ then $\vdash I_i\varphi$
G[≈]	If $\vdash \varphi$ then $\vdash [\approx]_i\varphi$
G[∘+]	If $\vdash \varphi$ then $\vdash [\circ^+]_i\varphi$

Table 3: Axiom schemata and inference rules of **AIL**

5.1 Soundness

We prove that every theorem of **AIL** is valid, which is the soundness of the system.

Theorem 3. *If* $\vdash \varphi$*, then* $\models \varphi$.

Proof. We prove it by induction on the structure of **AIL**. First, we prove that all the axioms are valid. For the I_i and $[\approx]_i$ operators, those axioms can be proved in a way similar to **S5** because both are necessity operators on equivalence relations.

- Since the axioms AC, AA, AI, A[$\approx$], A[$\circ^+$], and AE can be proved in a way similar to AN, we show only the proof for AN here. Suppose that $M, w \models A_i\varphi$, then $At(\varphi) \subseteq \mathscr{A}_i(w)$. Since $At(\varphi) = At(\neg\varphi)$, $At(\neg\varphi) \subseteq \mathscr{A}_i(w)$. Thus, $M, w \models A_i\neg\varphi$. The converse direction can be proved similarly.

- Since the axiom IA can be proved in a way similar to INA, we show only INA. Suppose that $M, w \models \neg A_i\varphi$, then $At(\neg\varphi) \subseteq \mathscr{A}_i(w)$. An awareness set is constant within both equivalence classes defined by $\sim_i$ and $\approx_i$, respectively, from the definitions. For all v such that $(w, v) \in \sim_i$, $At(\neg\varphi) \subseteq \mathscr{A}_i(v)$ and thus $M, w \models I_i\neg A_i\varphi$.

- For AA[$\approx$], suppose that $M, w \models A_ip \wedge p$, then $p \in \mathscr{A}_i(w)$ and $w \in V(p)$. For v such that $(w, v) \in \approx_i$, $p \in \mathscr{A}_i(v)$ and $v \in V(p)$ from the definition of the relation. Thus, $M, w \models [\approx]_ip$.

- For $\mathrm{K}_{[\circ^+]}$, suppose that $M, w \models [\circ^+]_i(\varphi \to \psi)$ and $M, w \models [\circ^+]_i\varphi$. For all v such that $(w, v) \in (\sim_i \circ \approx_i)^+$, $M, v \models \varphi \to \psi$ and $M, v \models \varphi$, then $M, v \models \psi$. Thus, $M, w \models [\circ^+]_i\psi$.

- For MIX, suppose that $M, w \models [\circ^+]_i\varphi$. Since $(\sim_i \circ \approx_i)^+$ is an equivalence relation, $M, w \models \varphi \wedge [\approx]_iI_i[\circ^+]_i\varphi$.

- For IND, suppose that $M, w \models [\circ^+]_i(\varphi \to [\approx]_iI_i\varphi)$ and $M, w \models \varphi$. For all v such that $(w, v) \in (\sim_i \circ \approx_i)^+$, $M, v \models \varphi \to [\approx]_iI_i\varphi$. Since this closure is an equivalence relation, $M, w \models [\approx]_iI_i\varphi$. The formula φ holds at all worlds from w on $\sim_i \circ \approx_i$, and $[\approx]_iI_i\varphi$ holds at those worlds as well. Therefore, $M, w \models [\circ^+]_i\varphi$.

The remaining task is to prove that if the antecedent is valid, the consequence is also valid for each inference rule. All of them are straightforward. □

5.2 Completeness

We prove the converse direction: every valid formula is a theorem of the system. The logic $\mathcal{AIL}$ is no longer compact because we can take a set of formulae such as $\Phi = \{([\approx]_iI_i)^n\varphi \mid n \in \mathbb{N}\} \cup \{\neg[\circ^+]_i\varphi\}$, where $([\approx]_iI_i)^n$ is n iterations of $[\approx]_iI_i$. Therefore, we restrict elements of maximal consistent sets to finite sets and then define a canonical model. This technique is employed in the proof on a logic with common knowledge, which is defined by the reflexive-transitive closure [30]. We customize the technique for our logic, similar to [20].

As the first step, we define a *closure* as a restricted set of formulae.

Definition 9. *Let $cl : \mathcal{L}_{AIL} \to 2^{\mathcal{L}_{AIL}}$ be the function such that, for every $\varphi \in \mathcal{L}_{AIL}$, $cl(\varphi)$ is the smallest set satisfying that:*

1. $\varphi \in cl(\varphi)$;
2. *If $\psi \in cl(\varphi)$, then $sub(\psi) \subseteq cl(\varphi)$, where $sub(\psi)$ is the set of subformulae of ψ;*
3. *If $\psi \in cl(\varphi)$ and ψ is not a form of negation, then $\neg\psi \in cl(\varphi)$;*
4. *If $A_i\psi \in cl(\varphi)$, then $A_i\chi \in cl(\varphi)$, where $\chi \in sub(\psi)$;*
5. *If $A_i\psi \in cl(\varphi)$, then $I_iA_i\psi, I_i\neg A_i\psi, [\approx]_ip \in cl(\varphi)$, where p is an atomic proposition in $sub(\psi)$;*
6. *If $I_i\psi \in cl(\varphi)$ and ψ is a form of neither $I_i\chi$ nor $\neg I_i\chi$, then $I_iI_i\psi$, $I_i\neg I_i\psi \in cl(\varphi)$;*
7. *If $[\approx]_i\psi \in cl(\varphi)$ and ψ is a form of neither $[\approx]_i\chi$ nor $\neg[\approx]_i\chi$, then $[\approx]_i[\approx]_i\psi$, $[\approx]_i\neg[\approx]_i\psi \in cl(\varphi)$;*
8. *If $[\circ^+]_i\psi \in cl(\varphi)$, then $[\approx]_iI_i[\circ^+]_i\psi \in cl(\varphi)$;*
9. *If $E_i\psi \in cl(\varphi)$, then $A_i\psi$, $[\circ^+]_i\psi \in cl(\varphi)$.*

Lemma 5. *For every φ, $cl(\varphi)$ is finite.*

Proof. We prove it by induction on the structure of $cl(\varphi)$. Since the number of subformulae of each formula is finite, all the conditions preserve the finiteness. □

We proceed to define a consistent set and a maximal consistent set in a closure. A set of the latter is used as a domain of a canonical model.

Definition 10. *Let Φ be the closure of a formula. Γ is a consistent set in Φ iff*

- $\Gamma \subseteq \Phi$,
- $\Gamma \nvdash \bot$.

Moreover, Γ is a maximal consistent set in Φ, iff Γ is a consistent set and

- *There is no $\Gamma' \subseteq \Phi$ such that $\Gamma \subset \Gamma'$ and $\Gamma' \nvdash \bot$.*

A consistent set can always expand to a maximal consistent set that includes the original set.

Lemma 6. *Let Φ be the closure of a formula. If Γ is a consistent set in Φ, then there exists a maximal consistent set Δ in Φ such that $\Gamma \subseteq \Delta$.*

Proof. It can be proved in the same way as Lindenbaum's lemma. We obtain a maximal consistent set by adding formulae in Φ to Γ so as to preserve the consistency. □

Next, we define a canonical model restricted by the closure of a formula and an equivalence relation depending on the model.

Definition 11. *Let Φ be the closure of a formula. A canonical model M^* for Φ is a tuple $\langle W^*, \{(\sim_i)^*, (\mathscr{A}_i)^*\}_{i \in \mathcal{G}}, V^* \rangle$, where:*

- $W^* := \{\Gamma \mid \Gamma$ *is a maximal consistent set in* $\Phi\}$;
- $(\Gamma, \Delta) \in (\sim_i)^*$ *iff* $\{\varphi \mid I_i \varphi \in \Gamma\} \subseteq \Delta$;
- $V^*(p) := \{\Gamma \mid p \in \Gamma\}$;
- $(\mathscr{A}_i)^*(\Gamma) := \{p \mid A_i p \in \Gamma\}$.

Definition 12. *Let Φ be the closure of a formula. A relation $(\approx_i)^*$ on W^* is defined by $(\Gamma, \Delta) \in (\approx_i)^*$ iff $\{\varphi \mid [\approx]_i \varphi \in \Gamma\} \subseteq \Delta$.*

We now prove the canonicity: a canonical model for a closure is an epistemic model with awareness.

Lemma 7. *For every φ, a canonical model for the closure of φ is an epistemic model with awareness.*

Proof. We prove that a canonical model for the closure of φ satisfies the definition of an epistemic model with awareness. For W^* and V^*, the proofs are trivial.

- For $(\sim_i)^*$, it is proved in a way similar to **S5**. For the reflexivity, it is sufficient to prove $(w, w) \in (\sim_i)^*$ for all $w \in W^*$. This follows from $\mathrm{T_I}$. For the symmetricity, it is sufficient to prove that for all $w, v \in W$, if $(w, v) \in (\sim_i)^*$, then $(v, w) \in (\sim_i)^*$. This follows from $\mathrm{T_I}$ and $\mathrm{5_I}$. For the transitivity, it is sufficient to prove that, for all $w, v, u \in W^*$, if $(w, v) \in (\sim_i)^*$ and $(v, u) \in (\sim_i)^*$, then $(w, u) \in (\sim_i)^*$. This follows through $I_i \varphi \to I_i I_i \varphi$ from $\mathrm{K_I}, \mathrm{T_I}, \mathrm{5_I}$, and GI.
- For $(\mathscr{A}_i)^*$, it is sufficient to prove that for all $w \in W^*$, if $(w, v) \in (\sim_i)^*$ then $(\mathscr{A}_i)^*(w) = (\mathscr{A}_i)^*(v)$. Suppose that $(w, v) \in (\sim_i)^*$ and $p \in (\mathscr{A}_i)^*(w)$. It follows from the definition of $(\mathscr{A}_i)^*(w)$ and IA that $I_i A_i p$. From the definition of $(\sim_i)^*$, $A_i p \in v$. Thus, $p \in (\mathscr{A}_i)^*(v)$. The converse direction can be proved from INA.

□

Lemma 8. *Relation $(\approx_i)^*$ on each canonical model is an A-equivalence relation on an epistemic model with awareness.*

Proof. First, we prove that $(\approx_i)^*$ is an equivalence relation. It can be proved in the same way as $(\sim_i)^*$ because both operators defining each relation have the axioms corresponding to K, T, and 5 axioms. Next, we prove that for all $(w, v) \in (\approx_i)^*$ and every $p \in (\mathscr{A}_i)^*(w)$, if $w \in V^*(p)$, then $v \in V^*(p)$ and vice versa. From left to right, suppose that $w \in V^*(p)$ for $p \in (\mathscr{A}_i)^*(w)$, then $p, A_i p \in w$. It follows that $p \in v$ from AA[$\approx$] and the definition of $(\approx_i)^*$. The converse direction can be proved by that $(\approx_i)^*$ is an equivalence relation. □

We prove a corollary of Lemma 6.

Lemma 9. *Let Φ be the closure of a formula, M^* the canonical model for Φ, and Γ a consistent set in Φ. For every $\varphi \in \Phi$, $\Gamma \vdash \varphi$ iff $\varphi \in \Delta$ for every maximal consistent set $\Delta \in W^*$ such that $\Gamma \subseteq \Delta$.*

Proof. From left to right, suppose that $\Gamma \vdash \varphi$, $\Gamma \subseteq \Delta$, and $\Delta \in W^*$. There exists a finite subset $\Gamma' \subseteq \Gamma$ such that $\vdash \bigwedge \Gamma' \to \varphi$. Since $\Gamma' \subseteq \Delta$, $\varphi \in \Delta$. From right to left, we prove it by contraposition. It is sufficient to prove that if $\Gamma \nvdash \varphi$, then there exists a maximal consistent set Δ such that $\Gamma \subseteq \Delta$, $\Delta \in W^*$, and $\varphi \notin \Delta$. By the assumption, $\Gamma \cup \{\neg\varphi\}$ is a consistent set in Φ. Thus, we obtain a maximal consistent set Δ in Φ such that $\Gamma \subseteq \Delta$ and $\varphi \notin \Delta$ by Lemma 6. □

The next task is to define a $[\circ^+]_i$-path and φ-path on the canonical model M^* for a closure.

Definition 13. *Let Φ be the closure of a formula and M^* the canonical model for Φ. A $[\circ^+]_i$-path from Γ is a sequence $\Gamma_0, \ldots, \Gamma_n$ of maximal consistent sets in W^* such that $(\Gamma_k, \Gamma_{k+1}) \in (\sim_i)^* \circ (\approx_i)^*$ for all k, where $0 \leq k \leq n$ and $\Gamma_0 = \Gamma$. The length of $\Gamma_0, \ldots, \Gamma_n$ is n. A φ-path is a sequence $\Gamma_0, \ldots, \Gamma_n$ of maximal consistent sets in W^* such that $\varphi \in \Gamma_k$ for all k, where $0 \leq k \leq n$.*

Lemma 10. *Let Φ be the closure of a formula, M^* the canonical model for Φ, and Γ and Δ maximal consistent sets in W^*. If $\bigwedge \Gamma \wedge \neg[\approx]_i I_i \neg \bigwedge \Delta$ is consistent, then $(\Gamma, \Delta) \in (\sim_i)^* \circ (\approx_i)^*$ in the canonical model M^*.*

Proof. It is sufficient to prove that if $\bigwedge \Gamma \wedge \neg[\approx]_i I_i \neg \bigwedge \Delta$ is consistent and $[\approx]_i I_i \varphi \in \Gamma$, $\varphi \in \Delta$. We prove it by the contradiction. From the assumptions, $[\approx]_i I_i \varphi \wedge \neg[\approx]_i I_i \neg \bigwedge \Delta$ is consistent. Suppose that $\varphi \notin \Delta$, then $\neg\varphi \in \Delta$. It follows that $[\approx]_i I_i \varphi \wedge \neg[\approx]_i I_i \varphi$ is consistent. This is a contradiction. Thus, $\varphi \in \Delta$. □

Lemma 11. *Let Φ be the closure of a formula and M^* the canonical model for Φ. If $[\circ^+]_i\varphi \in \Phi$ and $\Gamma \in W^*$, then $[\circ^+]_i\varphi \in \Gamma$ iff every $[\circ^+]_i$-path from Γ is a φ-path and $[\circ^+]_i\varphi$-path.*

Proof. ($\Rightarrow$) We prove it by induction on the length of a $[\circ^+]_i$-path.

- For the base case, suppose that the length of a $[\circ^+]_i$-path is 0, $[\circ^+]_i\varphi \in \Phi$, and $[\circ^+]_i\varphi \in \Gamma$. $\Gamma = \Gamma_0 = \Gamma_n$. By MIX, $\varphi \in \Gamma$.
- For induction steps, suppose that the length of a $[\circ^+]_i$-path is $k+1$, $[\circ^+]_i\varphi \in \Phi$, and $[\circ^+]_i\varphi \in \Gamma$. By the induction hypothesis, $[\circ^+]_i\varphi \in \Gamma_k$. It follows that $\varphi, [\circ^+]_i\varphi \in \Gamma_{k+1}$ from MIX and the definitions of $(\sim_i)^*$ and $(\approx_i)^*$.

($\Leftarrow$) Let $S([\circ^+]_i, \varphi)$ be a set of maximal consistent sets Δ in W^* such that every $[\circ^+]_i$-path from Δ is a φ-path. We introduce the following formula:

$$\chi = \bigvee_{\Delta \in S([\circ^+]_i, \varphi)} \bigwedge \Delta.$$

Suppose that every $[\circ^+]_i$-path from Γ is a φ-path. As the first step, we prove the following:

$$(1)\ \vdash \bigwedge \Gamma \to \chi; \quad (2)\ \vdash \chi \to \varphi; \quad (3)\ \vdash \chi \to [\approx]_i I_i \chi.$$

- For (1), since $\Gamma \in S([\circ^+]_i, \varphi)$, $\vdash \bigwedge \Gamma \to \chi$.
- For (2), $\varphi \in \Delta$ for every $\Delta \in S([\circ^+]_i, \varphi)$ since every $[\circ^+]_i$-path from Δ is a φ-path. Thus, φ is derived from χ.
- For (3), we prove it by contradiction. Suppose $\chi \wedge \neg[\approx]_i I_i \chi$ is consistent, then there exists Δ such that $\bigwedge \Delta \wedge \neg[\approx]_i I_i \chi$ is consistent because of the construction of χ. A W^* can be regarded as the whole set of combinations satisfying a condition of formulae in Φ, and χ as a representation of a particular set of the combinations. The formula $\neg \bigvee_{\Lambda \in W^* \setminus S([\circ^+]_i, \varphi)} \bigwedge \Lambda$ is equivalent to χ because a particular set of combinations represented by χ is the complement of the set containing all the other combinations. Therefore, $\bigwedge \Delta \wedge \neg[\approx]_i I_i \neg \bigvee_{\Lambda \in W^* \setminus S([\circ^+]_i, \varphi)} \bigwedge \Lambda$ is consistent. There exists Λ such that $\bigwedge \Delta \wedge \neg[\approx]_i I_i \neg \bigwedge \Lambda$ is consistent. By Lemma 10, $(\Delta, \Lambda) \in (\sim_i)^* \circ (\approx_i)^*$ and there exists a $[\circ^+]_i$-path from Δ but not a φ-path. This is a contradiction. Thus, $\vdash \chi \to [\approx]_i I_i \chi$.

By (3) and G$[\circ^+]$, $\vdash [\circ^+]_i(\chi \to [\approx]_i I_i \chi)$. It follows that $\vdash \chi \to [\circ^+]_i \chi$ from IND. By (1), $\vdash \bigwedge \Gamma \to [\circ^+]_i \chi$. By (2), $G[\circ^+]$, and $K_{[\circ^+]}$, $\vdash \bigwedge \Gamma \to [\circ^+]_i \varphi$. Thus, $[\circ^+]_i \varphi \in \Gamma$. □

We proceed to prove the truth lemma: in a canonical model, a true formula at a world is included in that world.

Lemma 12. *Let Φ be the closure of a formula and M^* the canonical model for Φ. For all $w \in W^*$ and every $\varphi \in \Phi$, $M^*, w \vDash \varphi$ iff $\varphi \in w$.*

Proof. We prove this by induction on the structure of formulae. The proofs of the base case and the logical connectives are straightforward.

- For the case of $A_i\psi$, we prove it by induction on the structure of ψ.
  - For the base case, suppose that $M^*, w \vDash A_i p$, then $p \in \{q \mid A_i q \in w\}$ and $A_i p \in w$. As for the converse direction, suppose that $A_i p \in w$. This means that $(\mathscr{A}_i)^*(w)$ contains p. Thus, $M^*, w \vDash A_i p$.
  - For the other cases, we obtain the desired proof by induction hypothesis and decomposing the formula using corresponding axioms: AN, AC, AA, AI, A[$\approx$], A[$\circ^+$], and AE. We demonstrate the proof for $A_i\neg\chi$.
    - Suppose that $M^*, w \vDash A_i\neg\chi$, which means that $At(\neg\chi) \subseteq \{q \mid A_i q \in w\}$. Since $At(\neg\chi) = At(\chi)$, $M^*, w \vDash A_i\chi$. It follows that $A_i\neg\chi \in w$ from the induction hypothesis and AN. The converse direction can be proved similarly.

- For the case of $I_i\psi$ and $[\approx]_i\psi$, it is proved in a strategy similar to **S5**. We only demonstrate the case of $I_i\psi$ here. From left to right, suppose that $M^*, w \vDash I_i\psi$, which means that for all v such that $\{\chi \mid I_i\chi \in w\} \subseteq v$, $M^*, v \vDash \psi$. It follows from the induction hypothesis that for all v including w itself, $\psi \in v$. By Lemma 9, $\{\chi \mid I_i\chi \in w\} \vdash \psi$. There exists a finite subset $\{\chi_1, \ldots, \chi_n\}$ of $\{\chi \mid I_i\chi \in w\}$ such that $\vdash (\chi_1 \wedge \cdots \wedge \chi_n) \rightarrow \varphi$. It follows that $\vdash (I_i\chi_1 \wedge \cdots \wedge I_i\chi_n) \rightarrow I_i\varphi$ from GI and K_I. Since a set $\{I_i\chi_1, \ldots, I_i\chi_n\}$ is included in w, $w \vdash I_i\psi$. Thus, $I_i\psi \in w$. The converse direction is straightforward.

- For the case of $[\circ^+]_i\psi$, suppose that $M^*, w \vDash [\circ^+]_i\psi$, which means that for all v such that $(w, v) \in ((\sim_i)^* \circ (\approx_i)^*)^+$, $M^*, v \vDash \psi$. This means that every $[\circ^+]_i$-path from w is a ψ-path and $[\circ^+]_i\psi$-path. By Lemma 11, $[\circ^+]_i\psi \in w$. Similarly, the converse direction can be proved.

- For the case of $E_i\psi$, it is proved by decomposing it into $A_i\psi$ and $[\circ^+]_i\psi$. Suppose that $M^*, w \vDash E_i\psi$, then $M^*, w \vDash A_i\psi$ and $M^*, w \vDash [\circ^+]_i\psi$. It follows from the proofs in terms of A_i and $[\circ^+]_i$ that $A_i\psi \in w$ and $[\circ^+]_i\psi \in w$. By EA[$\circ^+$], $E_i\psi \in w$. Similarly, the converse direction can be proved.

□

Lemma 13. *Let Φ be the closure of a formula and Γ a maximal consistent set in Φ. For every $\varphi \in \Phi$, if $\varphi \in \Gamma$ for every maximal consistent set Γ in Φ, then $\vdash \varphi$.*

Proof. We prove it by contraposition. Suppose $\nvdash \varphi$, then $\{\neg\varphi\}$ is a consistent set in Φ. By Lemma 6, there exists a maximal consistent set in Φ such that the set does not contain φ. □

Finally, we prove the completeness theorem.

Theorem 4. *For every $\varphi \in \mathcal{L}_{AIL}$, if $\models \varphi$, then $\vdash \varphi$.*

Proof. Suppose that $\models \varphi$. For the canonical model M^* for the closure of φ, $M^* \models \varphi$ by Lemma 7 and Lemma 8. Thus, for every maximal consistent set w in W^*, $\varphi \in w$ by Lemma 12. Therefore, $\vdash \varphi$ by Lemma 13. □

6 Related Work

There have been studies based on a similar idea to this paper, which is to associate an agent's awareness with indistinguishability among possible worlds or states.

Bisimulation In [27, 28, 29], the authors provided *awareness bisimulation* and proposed the notion of *speculative knowledge*. The bisimulation relation $\mathfrak{R}[\mathscr{A}(w)]$ is between the models that cannot be distinguished by formulae consisting only of aware atomic propositions. Speculative knowledge is defined in terms of the awareness bisimulation; formally, φ is speculatively known ($K^S\varphi$) at world w if and only if φ is true in all worlds that are $\mathscr{A}(w)$ awareness bisimilar to some accessible worlds from w. The notion captures a true proposition in all accessible worlds for every possible interpretation of all propositions of which an agent is unaware. As with our logic, their logic thereby formalizes indistinguishability among possible worlds induced by a lack of awareness.

However, this knowledge differs from that referred to by any of our operators. At w in M illustrated in Figure 9, K^Sp is true. This is because the truth value of the aware proposition p is the same in all worlds that are $\{p\}$ awareness bisimilar to w, such as w'. In contrast, $\neg[\circ^+]p$ and $\neg Ep$ are true at w in M. The observed difference is attributed to how to capture the notion of indistinguishability dependent on awareness/unawareness, that is, whether to capture awareness-indistinguishability at the model level or the world level. The formal comparative analysis is left for future work.

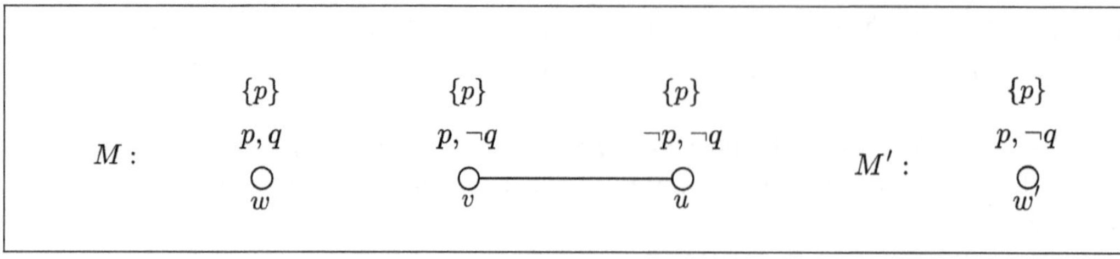

Figure 9: Possible world models with awareness for a single agent, where a solid line represents an IK-accessibility relation. Pointed model (M, w) is $\mathscr{A}(w)$ awareness bisimilar to (M', w'). For readability, we omit the reflexivity.

Equivalence relations among worlds In [20], the authors defined an equivalence relation expressing awareness-indistinguishability and, as with us, proposed a different definition of explicit knowledge from FH logic. Their approach inspires us to define our logic. They addressed a notion of awareness from an agent's perspective, that is, what "agent i is aware of a proposition from another agent j's perspective" is, representing it as $A^i_j\varphi$. We may construe the notion as information concerning another agent's awareness that an agent can use to obtain her explicit knowledge, and then represent it as explicit knowledge of awareness. Under this reading, the formula can be represented semantically within an epistemic model with awareness. Their logic does not handle a chain of perspectives with a depth of more than three, whereas our logic handles the chain with a finite but arbitrary depth. Furthermore, their semantics adopted a *global definition* of awareness sets; hence, agents' awareness is implicitly known to every agent, that is, for all $w \in W$, $M, w \models I_i A^j_k$ for each $i, j, k \in \mathcal{G}$. In contrast, our models adopt a *local definition*, which is one of the differences from their models.

In [24], the authors also introduced an equivalence relation between worlds sharing the same truth value for aware propositions. Adopting the global definition of awareness, they proposed *static abstraction operator* that refers to equivalence classes representing an 'abstraction' of the given model, which operator is the global counterpart of $[\approx]$ in $\mathcal{AIL}$.

Besides these awareness logics, some studies introduced the same type of equivalence relations. The papers [11, 12] proposed logics formalizing *ceteris paribus* clauses, which come from the philosophy of science and are understood in the sense 'all other things being equal.' The logics have the operator $[X]$ referring to a formula that holds in all worlds sharing the same valuation over a given set X of atomic propositions. The operator works the same as $[\approx]_i$ from a formal aspect; $[\approx]_i\varphi = [\mathscr{A}_i]\varphi$. Furthermore, the authors developed a translation of the $[X]$ operator into a universal operator, embedding their logics into the class of Kripke models with universal relations. This technique can be applied to our logic, provided awareness sets are finite.

Space-state model An *unawareness structure*, proposed in [16, 17], consists of a lattice of disjoint state-spaces with a partial order. Each space serves as a restriction on vocabulary instead of an awareness set. A state-space S greater than S' means that the states of S describe situations with a richer vocabulary than those of S'. The states in a space associated with an agent's awareness (and spaces associated with other agents' awareness from the agent's viewpoint, for multi-agent cases) can be interpreted as a subjective description of an epistemic situation in the agent's mind. This structure (henceforth, an HMS model) provides a formalization of awareness and explicit knowledge in a syntax-free way [3], having been applied to game theory [19, 18].

We can view partitions by an A-equivalence relation for each agent as another formalization of their spaces in various vocabularies. For instance, let p, q be atomic propositions and $\mathscr{A}_a(w) = \{p, q\}, \mathscr{A}_b(w) = \{p\}, \mathscr{A}_c(w) = \{q\}$, and $\mathscr{A}_d(w) = \emptyset$ for all w. For brevity, let us represent each world as a tuple of the literals; e.g., (p, q) represents the world in which both propositions are true. Their lattice comprises four state-spaces:

- $\{(p, q), (p, \neg q), (\neg p, q), (\neg p, \neg q)\}$;
- $\{(p), (\neg p)\}$;
- $\{(q), (\neg q)\}$;
- $\emptyset$,

whereas our model has four types of partitions:

- $\{[(p, q)]_{\approx_a}, [(p, \neg q)]_{\approx_a}, [(\neg p, q)]_{\approx_a}, [(\neg p, \neg q)]_{\approx_a}\}$;
- $\{[(p, q)]_{\approx_b}, [(\neg p, q)]_{\approx_b}\}$;
- $\{[(p, q)]_{\approx_c}, [(p, \neg q)]_{\approx_c}\}$;
- $\{[(p, q)]_{\approx_d}\}$.

A formal comparison would require a translation between IK- and EK-accessibility relations and their possibility correspondences [4]. Nevertheless, it appears that our model forms a single model that overlaps each space.

The studies [4, 3] that explored the relation between HMS models and FH logic showed that the explicit knowledge was equivalent to the intersection of implicit knowledge and awareness, as with FH logic. However, given the structural similarity between an HMS model and our model, their model appears to provide rich structures enough to interpret our operators. Furthermore, the structures proposed in [4, 3] are composed of multiple Kripke models (the former: the category of Kripke models, the latter: a Kripke lattice model), having been designed to correspond to the lattice of spaces in an HMS model. A formal comparison with these logics elucidates the knowledge that HMS models and $\mathcal{AIL}$ capture.

7 Conclusion

In this paper, we proposed a notion of indistiguishability among possible worlds due to a lack of awareness; that is, awareness-indistinguishability, which the standard accessibility does not capture. We demonstrated, using a concrete example, that the definition of explicit knowledge in FH logic ($E\varphi \leftrightarrow I\varphi \wedge A\varphi$) may derive undesirable propositions that cannot be considered explicit knowledge, and that the proposed indistinguishability plays an important role in the representation of the knowledge. Therefore, we focused on this indistinguishability and revisited the definition of explicit knowledge, proposing the logic $\mathcal{AIL}$ for representing the knowledge that is a refinement of that in FH logic. The indistinguishability is formally expressed by an equivalence relation between worlds depending on an agent's awareness, called A-equivalence relation. In our semantics, the aware implicit knowledge does not necessarily entail explicit knowledge ($E\varphi \rightarrow I\varphi \wedge A\varphi$ but $I\varphi \wedge A\varphi \not\rightarrow E\varphi$), which enables us to rule out the undesirable propositions. We then proved that our logical language is more expressive than that of an FH logic (Theorem 1) and showed that the latter is embeddable into $\mathcal{AIL}$ (Theorem 2). Furthermore, we provided the axiomatic system **AIL** and proved its soundness and completeness (Theorems 3 and 4).

There are several directions for further research.

Dynamics One important direction is to explore the information referred to by $[\approx]$ and $[\circ^{+}]$ more thoroughly. As we explored what kind of knowledge I and $[\circ^{+}]$ capture with dynamic operators in Subsection 3.3, the introduction of them [25, 26, 9] sharpens our understanding of these notions. Our definition of A-equivalence relation is model-dependent, which facilitates its extension into a logic involving dynamics.

Group knowledge Another direction is to extend our framework to group knowledge. Incorporating the notion of common knowledge is especially of interest. In strategic situations where a player makes a decision with knowledge of the other's action, it is valuable to represent common knowledge from the viewpoint of an outside observer with complete awareness. At the same time, it would be equally valuable to represent common knowledge from the viewpoint of an agent with incomplete awareness. Our framework could be applied to formalize such common knowledge.

Moreover, clarifying the relations with the logics mentioned in Section 6 would also be a promising strategy for gaining a deeper understanding of what those logics and our logic capture.

References

[1] Thomas Ågotnes and Natasha Alechina. A logic for reasoning about knowledge of unawareness. *Journal of Logic, Language and Information*, 23(2):197–217, 2014.

[2] Philippe Balbiani, David Fernández-Duque, and Emiliano Lorini. A logical theory of belief dynamics for resource-bounded agents. In *15th International Joint Conference on Autonomous Agents and Multiagent Systems (AAMAS 2016)*, pages 644–652, 2016.

[3] Gaia Belardinelli and Rasmus K. Rendsvig. Awareness logic: Kripke lattices as a middle ground between syntactic and semantic models. *J. Log. Comput.*, 33(6):1186–1215, 2023.

[4] Gaia Belardinelli and Burkhard C. Schipper. Implicit knowledge in unawareness structures. *Synthese*, 204(5):141, 2024.

[5] Michael Bergmann. *Justification without awareness: A defense of epistemic externalism.* Clarendon Press, 2006.

[6] Brian F. Chellas. *Modal logic: an introduction.* Cambridge University Press, 1980.

[7] Ronald Fagin and Joseph Y. Halpern. Belief, awareness, and limited reasoning. *Artificial Intelligence*, 34:39–76, 1988.

[8] Ronald Fagin, Joseph Y. Halpern, Yoram Moses, and Moshe Y. Vardi. *Reasoning About Knowledge.* MIT Press, 1995.

[9] Claudia Fernández-Fernández. *Awareness in Logic and Epistemology: A Conceptual Schema and Logical Study of the Underlying Main Epistemic Concepts.* Springer, 2021.

[10] Claudia Fernández-Fernández and Fernando R. Velázquez-Quesada. Awareness of and awareness that: their combination and dynamics. *Logic Journal of the IGPL*, 29(4):601–626, 2021.

[11] Davide Grossi. A note on brute vs. institutional facts. In *Normative Multi-Agent Systems, 15.03. - 20.03.2009*, volume 09121 of *Dagstuhl Seminar Proceedings*, 2009.

[12] Davide Grossi, Emiliano Lorini, and François Schwarzentruber. The ceteris paribus structure of logics of game forms. *Journal of Artificial Intelligence Research*, 53:91–126, 2015.

[13] Davide Grossi and Fernando R. Velázquez-Quesada. Syntactic awareness in logical dynamics. *Synthese*, 192(12):4071–4105, 2015.

[14] Joseph Y. Halpern. Alternative semantics for unawareness. *Games and Economic Behavior*, 37(2):321–339, 2001.

[15] Joseph Y. Halpern and Leandro C. Rêgo. Interactive unawareness revisited. *Games and Economic Behavior*, 62(1):232–262, 2008.

[16] Aviad Heifetz, Martin Meier, and Burkhard C. Schipper. Interactive unawareness. *Journal of Economic Theory*, 130(1):78–94, 2006.

[17] Aviad Heifetz, Martin Meier, and Burkhard C. Schipper. A canonical model for interactive unawareness. *Games and Economic Behavior*, 62:304–324, 2008.

[18] Aviad Heifetz, Martin Meier, and Burkhard C. Schipper. Dynamic unawareness and rationalizable behavior. *Games and Economic Behavior*, 81:50–68, 2013.

[19] Aviad Heifetz, Martin Meier, and Burkhard C. Schipper. Unawareness, beliefs, and speculative trade. *Games and Economic Behavior*, 77(1):100–121, 2013.

[20] Yudai Kubono, Teeradaj Racharak, and Satoshi Tojo. Logic of awareness in agent's reasoning. In *Proceedings of the 15th International Conference on Agents and Artificial Intelligence - Volume 1: ICAART*, pages 207–216, 2023. [Errata arXiv:2309.09214].

[21] Emiliano Lorini and Pengfei Song. Grounding awareness on belief bases. In *International Workshop on Dynamic Logic*, pages 170–186. Springer, 2020.

[22] Burkhard C. Schipper. Awareness. In Hans van Ditmarsch, Wiebe van der Hoek, Joseph Y. Halpern, and Barteld Kooi, editors, *Handbook of epistemic logic*. College Publications, 2015.

[23] Michael Sipser. *Introduction to the Theory of Computation*. Course Technology, second edition, 2006.

[24] Kosuke Udatsu and Katsuhiko Sano. Globally rigid awareness logic with static abstraction operators. In *International Workshop on Logical Aspects in Multi-Agent Systems and Strategic Reasoning (LAMAS&SR2024)*, 2024.

[25] Johan van Benthem. *Logical dynamics of information and interaction*. Cambridge University Press, 2011.

[26] Johan van Benthem and Fernando R. Velázquez-Quesada. The dynamics of awareness. *Synthese*, 177(1):5–27, 2010.

[27] Hans van Ditmarsch and Tim French. Awareness and forgetting of facts and agents. In *2009 IEEE/WIC/ACM International Joint Conference on Web Intelligence and Intelligent Agent Technology*, pages 478–483, 2009.

[28] Hans van Ditmarsch and Tim French. Becoming aware of propositional variables. In *Logic and Its Applications*, pages 204–218, 2011.

[29] Hans van Ditmarsch, Tim French, Fernando R. Velázquez-Quesada, and Yì N. Wáng. Implicit, explicit and speculative knowledge. *Artificial Intelligence*, 256:35–67, 2018.

[30] Hans van Ditmarsch, Wiebe van der Hoek, and Barteld Kooi. *Dynamic Epistemic Logic*. Springer Science & Business Media, 2007.

[31] Timothy Williamson. Inexact knowledge. *Mind*, 101(402):217–242, 1992.

Received July 2025

Spaces of Argumentation and their Interaction: The Role of Media in Public Discussion. Two Case Studies

Dov Gabbay
King's College London, UK
dov.gabbay@kcl.ac.uk

Gabriella Pigozzi
Université Paris-Dauphine, PSL Research University, Lamsade, 75016 Paris, France
gabriella.pigozzi@lamsade.dauphine.fr

Juliette Rouchier
Centre Gilles Gaston Granger, Maison de la Recherche, 29 Avenue Robert Schuman, 13621 Aix-en-Provence Cedex 1, France
juliette.rouchier@univ-amu.fr

Abstract

This article analyses how distinct argumentation spaces – media, law, and science – interact and collide, using Liebeck v. McDonald's as a legal case study and the hydroxychloroquine COVID-19 debate as a case of scientific dispute. Building on these two case studies, this analysis shows that arguments that intervene in many public debates originate from and move across different spaces, which employ different criteria, rules of evidence, and modes of reasoning for evaluating arguments. Furthermore, our study shows media-driven simplification, reframing, and truncation of nuanced expert exchanges, producing polarized public perceptions and policy consequences. The key features we identified include sliding arguments, repetition-driven strength (clones), shifting proof standards, hedgehog arguments, and zombie information, with implications for formalizing real-world argumentative dynamics and improving cross-space dialogue.

We thank the anonymous reviewers for their valuable comments and suggestions. In particular, we are grateful to Thierry Simonelli for his extensive observations and thoughtful criticisms of the manuscript.

1 Introduction

It is not uncommon to encounter the adjective "post-truth" to characterize the discursive context of our time. Cambridge dictionary defines it as "relating to a situation in which people are more likely to accept an argument based on their emotions and beliefs, rather than one based on facts". It is not a precise definition, but it describes a cultural context in which collective consensus about reality is less homogeneous than it was a generation ago within the same community or country. The appearance of the word, according to the Oxford Dictionary, dates back to 1989 in the US, which means that this word emerged in the context of a very active mediatised world, composed of multiple private, non-coordinated actors that have to attract the audience, and where state media have lost their status of cultural reference that can fix a collective truth. The mediatic frenzy and the need to seduce are thus a new context, where the mediatic world tends to produce an autonomous logic. Here we discuss the intrusion of this mediatic logic into two other cognitive universes, which are also supposed to be autonomous as social fields [7]: law and science.

In formal models, agents are typically assumed to share the same 'logic', recognising the same argumentative method and approach to determining the truth. However, in real societies, different groups meet in public spaces and debate using their own logic, beliefs, information, values, intentions, and analytical capabilities. One sociological theory that describes this situation is field theory from [6], which is related to the notion of habitus: the shared cognitive and action habits within a group that recognise its members by their ability to follow the rules. One application is the notion of taste (and its opposite, disgust), which reveals itself as a powerful differentiating characteristic between groups. A group is supposed to be autonomous and produce hierarchy thanks to its own recognition processes: a great scientist is supposed to be defined only by reference to their scientific creativity and not, for example, to their ability to perform in a TV show [5]. One can translate this idea in argumentation terms: various groups may have different criteria for accepting or rejecting arguments or sequences of arguments, depending on the role of debate whitin their social dynamics.

In science, scientists seek relevant ways to describe reality that increase humanity's knowledge, and this aim constrains their argumentation methods by accepting any argument they agree with and not building a counterargument to it [30]. But one can also identify the same specificity in legal cases where two opposing interests are pitted against each other through references to a vast corpus, referred to as "law" in our countries: they are also situations where arguments are not rejected *a priori* but accepted to generate discussion, negation, or approval.

Of course, the notion of social field has limits, in that it applies mainly to developed societies where those independent fields emerged: it assumes distinct (heterogeneous) social worlds that can each develop their own social dynamics independently of others (autonomy). Whereas each individual usually belongs to different fields (and thus navigates among diverse references), the fields keep independent logic. This we can observe in our work, where we identify encounters between different social (and thus argumentative) spaces. One critical field in post-modern societies is the media, and it happens to collide with most other fields, as it has become the most central builder of perceived reality. It thus has its own way of reasoning that is also independent of others [27].

This paper aims to describe some effects that can be observed in the collision of different argumentation frames. In particular, we observe how the "mainstream media" (seen as a significant category composed of journalists who must transmit information to a large public) transform specific arguments. To do this, we describe two instances of collisions with the media space, each concerning a social space in which specific argumentation frames exist: the judicial world and a scientific dispute. In analysing two different discourses in these cases, we observe that the media world tends to truncate argumentative discussions, meaning that the size of a debate is artificially limited after a couple of exchanges. However, this is not the case in a trial, where each argument is addressed, or in science, where each argument can serve as the starting point for a new argument chain. This simplicity could be explained by the media's rapid temporality (news must be published rapidly and continuously), which makes it challenging to align with the inherent uncertainty of real disputes. Another possible explanation is that the translation from the specialized world to the mediatic world leads to an oversimplification of complex matters.

We explore these dynamics to identify salient features and, in future work, to produce a formal model of these interactions that reproduces the observed limitations of expression. Here, we already propose representations of dynamics and structures for the organization of these argumentative discussions.

Our analysis is based on two cases. The first case is the U.S. legal case Liebeck v. McDonald's Restaurants. The description is primarily based on a documentary in which a synthesis has already been presented. The second case is the debate around the efficacy of a molecule (hydroxychloroquine) during the COVID-19 pandemic. The synthesis of the latter case is our own and therefore draws on many different sources.

These cases differ in their temporal and geographical dimensions, as well as the contexts of the debates (legal vs. scientific). Despite the differences, these cases highlight similar media distortions that employ argumentative methods, such as the hedgehog argument, repeated arguments, and a lack of (or a severely reduced and oversimplified) opposition information.

These debates are represented as argument spaces in which arguments move, are selected and reframed by the media, thereby simplifying complex exchanges. In both cases, long, nuanced debates are reduced to binary arguments – hot coffee to trivial (coffee is, by definition, hot), and hydroxychloroquine to a miracle treatment. Despite the differences – legal vs. scientific – the cases share similarities. A key difference is the impact: the hot coffee case had little societal effect, whereas the HCQ debate influenced public policy and lives worldwide, with more argument exchanges crossing spaces.

What emerges is that, across different argumentation spaces (law, science, media, public policy, public, etc.), the way people discuss relies on different proof standards (what counts as a good argument, and how do we organize the exchange of arguments?). Arguments within a specific space may not make sense to those outside it, as the language, rules of evidence, and reasoning can differ. Understanding these dynamics is crucial for facilitating dialogue among different parties and increasing the likelihood of finding common ground. Moreover, from the perspective of argumentation theory, public controversies are an important object of study [18, 20]. Public debates and controversies reveal facts about argumentation practice that mainstream theory may overlook. Studying them enables us to align contemporary argumentation theory with practice and to uncover new phenomena (like novel reasoning standards) that expand the field's theory.

Argumentation theories are diverse and include formal, informal and computational argumentation. They aim to describe real-world argumentative phenomena and to study how to resolve potential disagreements. However, as discussed in [58, 59], there is a gap between argumentation theories and applications of argumentation. Our work is also a step toward connecting real-world argumentative phenomena to formal argumentation. The relevance to the field of formal argumentation lies in the exploration of the dynamics of argumentative interactions, which can be formalised and studied within argumentation theory.

2 The hot coffee case

2.1 The story

The Liebeck v. McDonald's Restaurants (Case No. D–202-CV–199302419), in the U.S. District Court for the District of New Mexico, commonly known as "the hot coffee lawsuit", occurred in 1992 when a 79-year-old woman suffered second and third-degree burns from a spilt cup of coffee at a McDonald's drive-thru. The plaintiff attempted to settle with McDonald's for $15,000-$20,000 and requested that the company change its policy on hot coffee, but the company offered $800 and refused to lower the temperature of its coffee. The plaintiff filed a lawsuit against McDonald's and won. The jury awarded the plaintiff $200,000 in compensatory damages and $2.9 million in punitive damages, which was later reduced to $480,000. The parties reportedly reached a confidential out-of-court settlement later.

The $2.9 million punitive damages award made headlines, and Liebeck's case sparked a debate over tort reform, leading to a clash between lawyers and politicians. Lawyers saw it as an example of the justice system ensuring that an individual can prevail against a giant corporation when it does something wrong. Politicians, on the other hand, viewed it as an abuse of legal power. Susan Saladoff's documentary *Hot Coffee* [49] reconstructs how Liebeck's case was discussed in the public sphere, with politicians shouting outrage and sitcom sketches portraying the perceived absurdity of a person considering a lawsuit because her coffee was hot.

The media's reconstruction of the case was often one-sided: it was said that Stella Liebeck was in the car and burned herself while attempting to open the lid. This led public opinion to believe that she was burned while driving, thereby making her liable for being inconsiderate with a beverage that, by definition, is hot. In reality, Stella Liebeck's grandson was driving the car when the incident occurred. He parked it, she attempted to add cream and sugar to her coffee, placed it between her knees, and removed the lid. Describing the accident as occurring simply as when she was in the car amounts to what in the literature on the different forms of dishonesty has been identified as *deception*, that is, the information given is correct but in such a way that the recipient will most likely draw an incorrect conclusion [28]. In a way, presenting only one aspect of the case can be seen as a form of manipulation.

Given that coffee is naturally hot, the trial revealed that McDonald's coffee machines were defective, producing coffee that was unreasonably hot and increasing the likelihood of severe burns compared to coffee from other establishments. During the trial, evidence was presented that McDonald's policy required coffee to be served at 180° F to 190° F (compared to 150° F in other restaurants), which could cause third-degree burns within three seconds. During the trial, Liebeck's surgeon, D. Arredondo, declared that if liquid at that temperature makes contact with skin for more than a few seconds:

> If you're lucky, it will produce second-degree burns. If you're not as lucky, you will get third-degree or full-thickness burns requiring skin grafting and surgery.

This sustained the claim that coffee was not just hot, but dangerously hot. McDonald's blamed Liebeck for not removing her clothes immediately (she was wearing sweatpants, which absorbed the coffee and kept it against her skin) also considered her age, since older people's skin is thinner and less resistant to injury.

It also emerged that McDonald's was aware of at least 700 prior hot coffee burns between 1983 and 1992. This was deemed statistically insignificant, considering the billions of cups of coffee sold annually, to justify a change in their policy. Attorney Tracy Jenks tried the case for McDonald's and declared:

> McDonald's had a really, really strong reason for why they brewed their coffee at the temperature they did. It was an industrial standard based on the maximum extraction of the flavor in the maximum holding temperature. [46]

Another (unofficial) explanation - not endorsed by McDonald's - circulated about the extreme coffee temperature. McDonald's offered unlimited free refills. Serving coffee at a very high temperature could have been used to reduce the number of free refills, since people had to wait too long before they could drink the beverage.

In Susan Saladoff's documentary [49], it is highlighted how after the verdict, Liebeck became the old lady who spilt coffee on herself and got millions, a joke on late-night shows and was used by politicians to push for a tort reform that would limit "frivolous lawsuits". In contrast, McDonald's had been cast in the public mind as the victim.

> The size of the award got the media's attention, but it overshadowed the rest of the story. Details of the case and the facts related to how the jury made its decision went mostly unreported. [49]

As observed by John Llewellyn, Professor of Communication at Wake Forest University:

> 697 words in the Albuquerque Journal became 349 words in the Associated Press and became as few as 48 words in various renderings by major metropolitan newspapers. 48 words can't explain a lot. And then woman, coffee, millions sounds like a ripoff, not like a logical consequence of a thoughtful trial. [46]

In formal models of argumentation, only new arguments can be added, and repetitions are forbidden. However, in real life, repeating an argument can strengthen it (in the literature, this phenomenon is known as the *validity effect* [45]). In the subsequent analysis, we will refer to repeated arguments as *clones*. At the time of the Hot Coffee documentary, 19 years after the events took place, Liebeck was still seen as the greedy old woman:

> Very much like urban legends, it is a very compelling story. Once everybody decides what is true about something and the media has been sort of an echo chamber for it, then how do you deal with the fact that they might be wrong? (John Llewellyn in [49])

2.2 Argument spaces

In 1958, Toulmin [54] distinguished between *field-invariant* and *field-dependent* elements in argumentation:

> How far, for instance, can one compare the standards of argument relevant in a court of law with those relevant when judging a paper in the Proceedings of the Royal Society, or those relevant to a mathematical proof or a prediction about the composition of a tennis team? ([54], p.15).

He drew parallels between the judicial process and rational argumentation, highlighting the variability of evidence in a civil or criminal case, for example, but the commonality of procedural phases. Argumentation across fields follows a fixed procedure, suggesting a field-invariant sequence of steps of presenting the problem, gathering evidence, and reaching a conclusion. On the other hand, the criteria used in evaluating the validity of argumentation are field-dependent.

The same year as the publication of Toulmin's influential book, Perelman and Olbrechts-Tyteca published *La nouvelle rhétorique: Traité de l'argumentation* [36] (an English translation appeared in 1969 [37]). Their theory of argumentation focused on its persuasive aspect:

> [S]ince argumentation aims at securing the adherence of those to whom it is addressed, it is, in its entirety, relative to the audience to be influenced. ([37], p. 19])

Unlike in formal logic, where the validity of an argument depends on the relation between the premises and the conclusion and must be compelling to everyone, argumentation in ordinary language is relative to the *audience*, defined as "the ensemble of those whom the speaker wishes to influence by his argumentation" ([37] p. 19). Perelman and Olbrechts-Tyteca distinguished between a *particular* and a *universal* audience: an argument is always directed to a particular individual or group of people. In contrast, the universal audience ideally includes all reasonable human beings, thereby capturing the ideal shared standards of what constitutes a good argument.

The shift in audience is striking in Liebeck's case. Once attorneys became involved, the arguments shifted from personal experience to legal implications. The change in the audience was parallel to the shift in the argumentation style and rules of the debate. Once the focus shifted to the legal notions of damage and responsibility, the language, rules, and methods of presenting arguments evolved to meet the specialized standards of the legal field. The evaluation of arguments also depends on the specific area in which they circulate, with each community following its *field-dependent* standards. At the same time, from a rhetorical point of view, once the case began to be discussed in a legal court, arguments were advanced and formulated to persuade a different audience than in the personal experience, as in Perelman and Olbrechts-Tyteca' approach.

In our analyses, to encapsulate Toulmin's field-dependent standards and Perelman and Olbrechts-Tyteca's particular audience, we will refer to *argument spaces*. Our notion of "argument spaces" recalls Goodnight's three types of argument spheres: personal, technical, and public [17][1]:

> "Sphere" denotes branches of activity— the grounds upon which arguments are built and the authorities to which arguers appeal. Differences among the three spheres are plausibly illustrated if we consider the differences between the standards for arguments among friends versus those for judgments of academic arguments versus those for judging political disputes. ([17], p.216)

While Toulmin's fields prioritize members' expertise, argument spheres are social constructs and focus on how the arguer, recipient, and situation intersect. In the personal sphere, individuals in a personal relationship make and evaluate the arguments. In the technical sphere, arguments are produced, evaluated, and circulated among experts in their respective fields. Finally, public sphere arguments are directed towards general audiences who cannot answer, thus they constitute a transfer of argument rather than a dispute. Among the three spheres, the technical sphere is the most tightly linked to fields.

Our analyses of these two cases (the hot coffee lawsuit and the debate over hydroxychloroquine in fighting COVID-19) may help illustrate the differences in the formulation, presentation, reception, and evaluation of arguments depending on the sphere in which they circulate. Moreover, these two cases share the circulation of arguments between different spheres. The study of this dynamic allows to witness some interesting phenomena (cf. Section 4).

[1]The hot coffee lawsuit case was used as an example of Goodnight's argument spheres in [19].

In the hot coffee lawsuit (and, as we shall see later, in the hydroxychloroquine case), the media played a crucial role in the transmission of arguments across spaces. The media are governed by their own criteria of what constitutes a good argument. In addition to being an argumentation sphere of discursive production, they also act as a filter, by selecting which information to pass from one space (technical, for example) to another (the public), and, as we will see, the information is *simplified* to this end. It can also happen that the piece of news is oversimplified, meaning that it can be qualified as an (involuntary) deception. This can happen for several reasons: the media must publish daily, and complex technical topics must be made comprehensible to the general public. Additionally, newspapers and TV channels often belong to media outlets with political affiliations, which impose a particular framing on news stories. Examples of oversimplification leading to deception are the reporting that Stella Liebeck was in the car when the hot coffee accident happened (letting the hearer believe that she was driving) or saying that the coffee was hot, a kind of tautological fact (omitting that hot is a fuzzy term that admits several grades, from not hot to dangerously hot).

The argumentation spaces in the hot coffee case and their interactions can be represented as in Figure 1. This illustrates how arguments and facts move from one space to another and are consequently reshaped: arguments grounded in the personal experience of being burned and in pain (thus supporting a compensation claim) became legal arguments when lawyers took the case to court (represented as arguments of different shades and shapes in Figure 1). The main feeding channel to the public space is ensured by the media, which, as we have seen, selects which news passes by and (out of necessity and/or political agenda) simplifies it.

We now turn to the second case study: the debate about hydroxychloroquine during the COVID-19 pandemic in France. The interest of that debate was to make visible the construction of scientific knowledge in real time, the massive presence of scientists in the media, and, as we shall see, the collision between science and the media.

3 The hydroxychloroquine case

During the COVID-19 pandemic, public policy decisions were made in an unusual context, involving novel actions such as lockdowns, curfews, and school closures. Public decision-makers had to justify their choices, leading to arguments circulating in public spaces. Two key characteristics emerged: most arguments were labelled as "scientific" to be accepted, and there was high polarisation and rejection of arguments, often accompanied by abrupt explanations, which were frequently justified

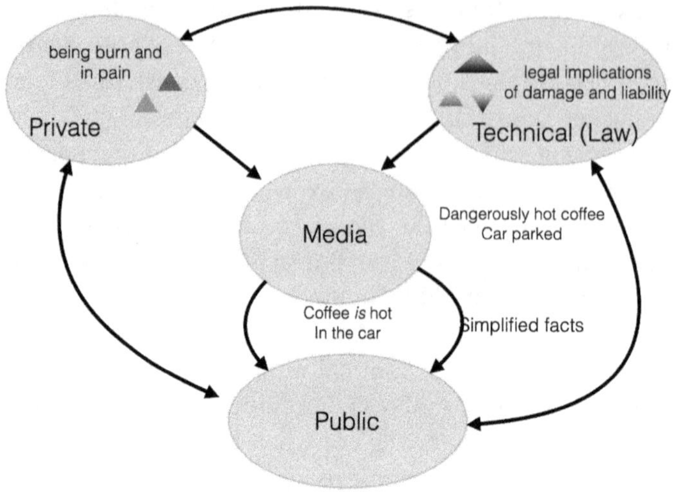

Figure 1: The Hot Coffee argument spaces

by *ad hominem* and *ad personam* attacks (as shown in our reconstruction in Figure 2).[2] Generalist media and scientific journals were the primary spaces for these interactions. We focus on early visible arguments but also briefly describe the evolution of the dispute over time.

After over five years of crisis, several public controversies remain:

- Early cures exist and are effective (HydroxyChloroquine, Azithromycin, Ivermectin, Vitamin D, Zinc, Steroids, Colchicine), though Remdesivir is now proven useless.
- The legitimate methods for proving a cure's efficiency (random control trials vs. observational history) are debated.
- The effectiveness of lockdowns on epidemic dynamics and mortality rates is questioned.
- The origin of the virus (human-made or natural) is uncertain.
- The safety of mass vaccination and potential dangers are concerns.

[2]The manual extraction of arguments to reconstruct a debate is a method used in argumentation, see for instance [39, 50].

We focus on the first two linked disputes: the potential use of Hydroxychloroquine (HCQ hereafter) in treating COVID-19 patients and its trustworthiness. We chose this debate because HCQ, a well-known medicine, sparked intense public debates with scientific justification. These debates divided families and friends, with HCQ becoming a symbol of a harsh scientific dispute. In fact, so harsh and emotional that two papers against the use of HCQ [29, 41] and one paper claiming its sufficient efficiency in an epidemiological context [15] were retracted.

3.1 Mediatic world asking for science never to mistake

The designation of "scientific" is vague, as [24] point out. Scientific production involves two phases: research and observation, which produce new knowledge, and academic reproduction, which transmits past knowledge. These phases employ different procedural methods and recognise distinct proof systems.

Furthermore, the concept of science is employed in various contexts, with the media often portraying it as an "objective truth" [21]. At the onset of the European crisis in February 2020, scientific knowledge was not yet complete, as the specific coronavirus had not been encountered six months earlier.

In a purely scientific world, agreement on reality emerges only after a long academic dispute. This dispute involves well-defined paradigms, methods, measures, and interpretations, with limits that define the truth of acquired knowledge. Transparency enables precise critics to show new data or interpret data differently. The first proposer responds, and the process continues. "Refutability" refers to the collective effort to promote transparency and open criticism, thereby revising proofs or analyses.

> Science and scientific objectivity do not (and cannot) result from the attempts of an individual scientist to be 'objective,' but from the friendly-hostile cooperation of many scientists. [40]

This discussion can be challenging, especially with frontier knowledge, as scientists often do not recognise the same paradigm [22]. Their assumptions differ, leading to incompatible methods and proof recognition. The charity principle [43] is crucial for a positive process: assume the other is rational and try to understand their reasoning rather than project your own. Even without shared pre-existing understanding, it is possible to follow someone else's reasoning and proof before criticizing it.

Unlike scientists, journalists must provide the public with information on a daily or weekly basis.

Two instruments guarantee the deontology of journalism. The first is that information must be linked to a verifiable source, such as data, people, or institutions. A serious article on a debate should cite sources from both sides to expose opposing arguments and beliefs. This aspect resembles scientific citation habits. However, as [2] observed, media reports on global warming do not reflect the scientific consensus due to the norm of "balanced" reporting, which gives undue weight to weak evidence and can increase confidence in erroneous claims.

The second is that a mistake is considered a sin, making the journalist unreliable for the following period. This idea does not distinguish between different ways of being wrong in science:

- Predicting is risky because models are based on assumptions, and the complex world often misleads them. This does not mean that models are always wrong; they may simply be unsuitable for prediction.

- Describing/explaining events can be wrong due to reductionist assumptions or missing explanatory elements. Mistakes benefit scientists by allowing them to refine the applicability of their models.

- Models can be proven wrong in many applications, prompting revisions. However, there is a lag between recognising a model's ineffectiveness and its eventual disappearance, making it difficult to determine whether the scientist or the theory is at fault. What is important for a scientist is thus not to avoid errors, but to be able to recognize the error and revise either her way of applying the model, or the theory itself. This humble attitude fosters trust among colleagues, which is essential for accepting the normality of errors. Bad faith, voluntary cheating, and various forms of default are possible in science. The media often expect scientists to be infallible, viewing science as a revelation rather than a journey towards truth. This religious relation leads to public excommunications and the belief in definitive truths.

3.2 What is an acceptable proof?

As mentioned in the Introduction, this paper focuses on the dispute in France regarding the effects of HCQ treatment on the SARS-CoV-2 virus. Here are some salient points of the debate (see Figures 2 and 3).

The first results announced by Didier Raoult (reduction of viral load in individuals after HCQ was administered and that the effect was even more striking if azithromycin (Az) was added) were refuted because these results came from a very small set of individuals, indicating a possible effect rather than proof. Raoult responded that, if a treatment was working, it was more important to care for the patients and let others demonstrate the treatment's efficacy. This aligns with a long-standing debate about ethics within the medical community [9].[3] When answering to Philippe Berta (MODEM MP) at the COVID-19 Commission of Inquiry (hearing of 24th June 2020), Didier Raoult replied:

> [My test] is the only way to make a comparison and make a significant difference. And contrary to what you say, and I apologise for this, but it is a radical error, the fewer people there are when it's significant, the more significant it is. When you need 10.000 people in a trial to be statistically significant, that means there's no difference. [...] When you're looking for 1% differences, you're no longer in medicine, you're in methodological fantasy. So any trial involving more than 1.000 people is just a trial trying to demonstrate something that doesn't exist. [...] Methodology is not science; it's a tool of science. [...] You need to read in medical books and in the field of infectious diseases how many therapies are currently being used that are based on randomised trials, and you'll see that there aren't many of them, and the few that there are – especially for AIDS – were totally useless because all you had to do was establish viral loads. So it's not true, it's a hoax, it will disappear just as it did twenty years ago. [...] All the people who have criticised this have only published comparative, non-randomised trials. [...]" ([44], our translation)

If Raoult was supported by some scientists or philosophers of science (such as [38, 53]) who expressed their agreement in the media, the opposing view was that a randomized controlled trial (RCT) is necessary to prove the treatment: the medicine has to be proven better than placebo, better than any other used molecule. Raoult then wrote a tribute to support his method, arguing that the reusability of an old molecule is a bet that should be made (and that the States should finance it). He also attacked the RCT on the basis that they are expensive experiments, thus requiring financing by the pharmaceutical industry, insinuating that they are certainly biased.

Later, he added the argument that the number of people who must be enrolled in an RCT for a disease that kills 0,5% poses serious ethical concerns [34]: 20.000 patients must be treated with the molecule and 20.000 without.

[3] One interesting argument: the RCT was established to get rid of mandarinate in medicine, and Raoult looks like the worst example of this mandarinate, so it is logical that he should be against the method.

Interestingly, the media regarded RCT as the "gold standard" from the start. However, researchers do not all agree on this status of ultimate proof: for example, [31] showed that one RCT cannot prove more than any other statistical study. Statistical proof is indeed a proof of non-rejection of a hypothesis, not a proof that the hypothesis is correct. Furthermore, there had already been a large discussion in France about the logic of using RCT during an epidemic, during the AIDS epidemic, and in 1994, a very large association of AIDS people succeeded in proving that conducting RCT is non-ethic as a huge number of people (randomly chosen) would be condemned while others would be given a promising therapy [3]. This information could be known in the media world but was not taken into account.

This mediatic discussion provoked a high polarization in most countries, where people seemed to perceive different realities that could hardly be made coherent: some asking for care (the Raoult side) and others asking for proof. This polarization seemed to have no end and evolved over time: while centered on a relatively straightforward medical question, it evolved into a never-ending battleground for argumentative debate.

Once Raoult (and his institution, IHU) was attacked, anyone who used their arguments was accused of being under his influence and was disqualified as irrational. This phenomenon was almost impossible to reconsider in time (still in 2025). To illustrate this, the Lancet paper case is very interesting: this scientific paper experienced a very swift retraction, but this did not alter policies or the actions of journalists.

3.3 Lancet or not Lancet?

A paper published in the prestigious medical journal The Lancet showed that the combined treatment with HCQ and AZ kills people [29]. This led to the halting of HCQ trials in the French trial, Discovery. The lead author (Dr Mandeep Mehra) was a renowned vascular surgeon at Harvard University. The study was based on the Surgisphere database, which included 96.000 patients' medical records from 671 hospitals in 11 countries from 20/12/2019 to 14/04/2020; 14.888 patients were treated with HCQ. The paper concluded that HCQ was ineffective for treating COVID-19 and was associated with increased mortality. Interestingly, the paper's results did not come from an RCT. But it was a meta-analysis. Still, it had massive consequences.

The Lancet paper was published on Friday, May 22nd 2020. Less than 24 hours later, the World Health Organization (WHO) stopped all 131 HCQ COVID-19 trials (including the European trial, Discovery and Solidarity), with thousands of patients already enrolled. Five days later, the use of HCQ to treat COVID-19 was forbidden in France.

The day before (May 26th 2020) in Italy, the AIFA (*Agenzia Italiana del Farmaco*, the Italian Medicines Agency) forbade the off-label use of HCQ. Forty-five doctors contested it, arguing that HCQ, based on clinical studies published in accredited international journals, can halt disease progression in early-stage patients and that they were denied their decision-making autonomy, as protected by the Italian Constitution. Seven months later, they won. The Consiglio di Stato's judges, Italy's highest administrative court, wrote:

> [O]nly if one of these conditions - efficacy or safety - were lacking, could AIFA legitimately suspend the off-label use of the drug to treat the virus and restrict the decision-making autonomy and prescriptive freedom of the doctor. ([1], p.4, our translation)

Therefore:

> [T]he persistent uncertainty as to its therapeutic efficacy, admitted by AIFA itself as justification for further evaluation in randomised clinical trials, is not sufficient reason in legal terms to justify the unreasonable suspension of its use in Italy by treating physicians based on a conclusion - the drug's *total definitive inefficacy* in every respect, including immunomodulatory - which, in the state of knowledge and research that is still partial and provisional, seems *radical and premature* even at a scientific level. ([1], p.9, our translation)

The Italian judges echoed one of Raoult's arguments that in an emergency, the priority is to treat patients, when they noted that:

> The application of the principles of evidence-based medicine in the face of an emergency situation that does not allow for the acquisition of certain and rapid experimental evidence from randomised and controlled studies on a reference population [...] leads to the paradox of denying any possibility of actually experimenting with treatment precisely when the need is greatest and most urgent for this class of patients, to avoid their hospitalisation and, in the most serious cases or comorbidities, death. ([1], p.9, our translation)

Two weeks after its publication, on 2 June 2020, The Lancet retracted the paper, described by its editor as "a shocking example of research misconduct". The fraud was so blatant that even people who disagreed with Raoult defended him. To mention just a few examples of evident flaws in the Lancet paper: all countries had the same percentage of smokers; in 2008, Surgisphere was a publisher of medical textbooks (how could it become the owner of such a powerful database?), Surgisphere had only 11 employees, and they had no scientific background [10].

Figure 3 reconstructs the circulation of arguments around the publication of the Lancet's paper. Although being the quickest ever retracted paper, journalists continued to consider that HCQ was dangerous (and, so, Raoult was a dangerous man), a phenomenon that could be labelled as an *information zombie* [33]. The same results as those reported in the Lancet paper were promoted in a publication by French authors in 2024. This was also revealed to be fraudulent and retracted in August 2024 [41]. Interestingly, although the Lancet paper had been retracted, the HCQ COVID-19 trials were not continued. So, the results never attained statistical significance.

It is worth noting that the attitude towards HCQ varied across countries, as illustrated in Figure 4.

All this discussion took place while papers were still being produced at a high rate and could be accessed quite easily (https://c19hcq.com). In most media discourses, people who still wondered whether (and how) HCQ could work against SARS-CoV-2 infections were judged silly. And yet, if so many papers got published by diverse researchers, it should mean that the dispute was not yet over.[4] These two worlds produced acceptable argument sets that are disjoint.

4 Some salient argumentation features

In this section, we describe some salient argumentation features that we have observed in our analyses.

4.1 Different notions of defeats or acceptable arguments

In a very general way, participants do not necessarily share the same notion of *defeat*. The media's view seems to be that there is a disagreement over two incompatible issues (T and T') between two parties A and B:

[4]On May 4th, 2023, Raoult wrote this tweet (our translation): "Let's talk science. I currently agree with the ANSM on hydroxychloroquine against Covid. It no longer works against the 2022 Omicron variant, as the concentrations required to inhibit Omicron are 50 times higher than those for other variants. The concentration of hydroxychloroquine in the blood using our protocol was three times the effective dose for the old variants. This is science, not dogma."

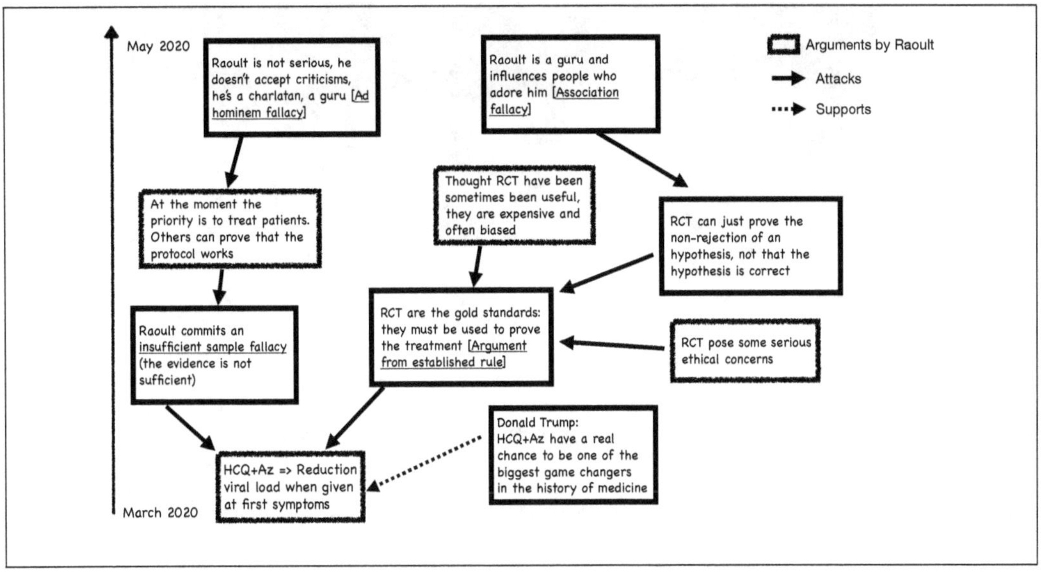

Figure 2: Debate in March-May 2020

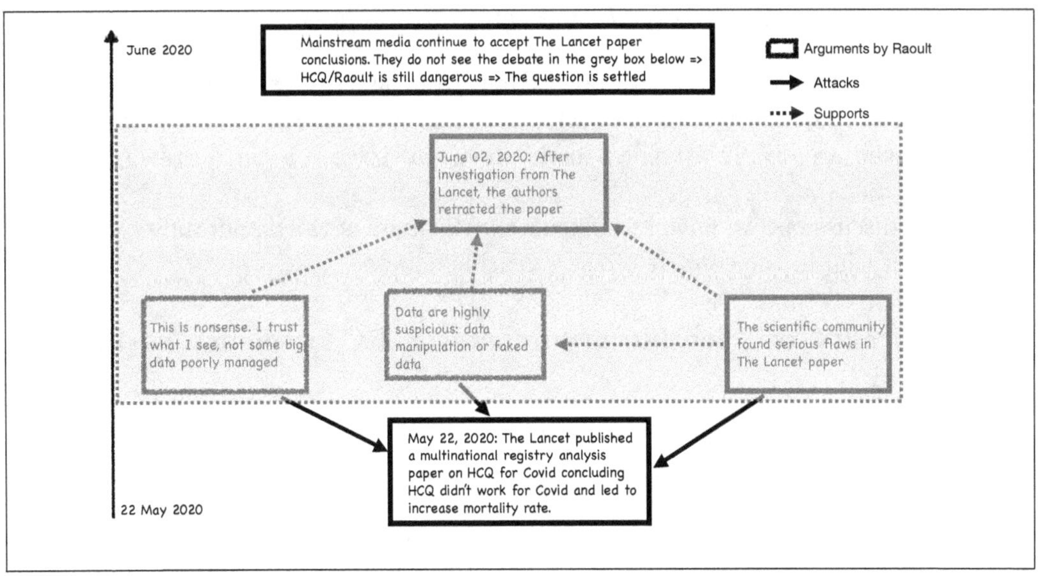

Figure 3: The Lancet paper (May-June 2020)

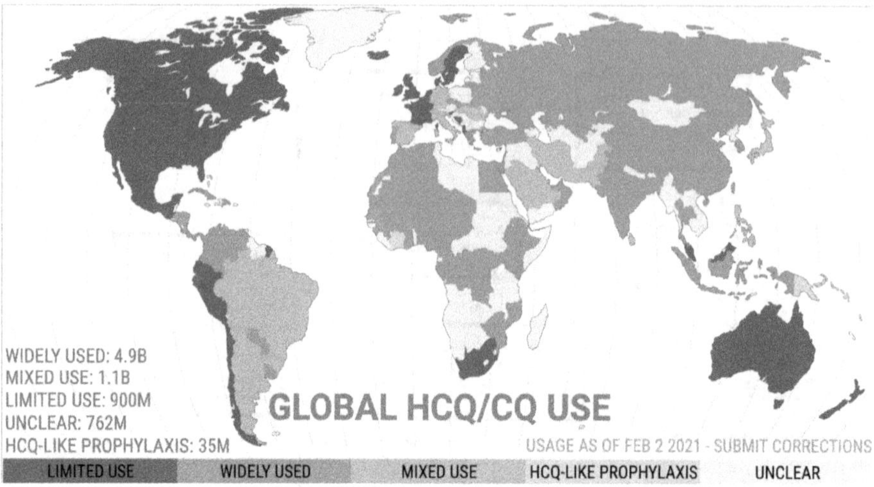

Figure 4: Use of HCQ in different countries in 2020. Source: https://c19early.org

$$A \text{ wins over } B = A \text{ proves } T \text{ or falsifies } T'$$

Whereas the scientific view is that:

$$A \text{ wins over } B = A \text{ refutes } B\text{'s arguments}$$

Interestingly, the two views above come from [35], where Pera compares the views of science of the philosopher-scientists who gave rise to modern science (the Fathers), who are deeply attached to the idea that science is uncontroversial, with ours (their Sons).

In the discussions we have reconstructed, the issue of the disputability of knowledge was crucial in the context of the scientific debate.

4.2 Plurality of scientific arguments or RCT: science vs bureaucracy?

The debate over HCQ highlighted two competing notions of what constitutes good evidence. According to [25], on the one hand, there are public health epidemiologists who are "methodologically liberal and pragmatic" and accept using diverse kinds of data. On the other hand, the clinical epidemiologists, who call for an evidence-based medicine, "gold standard evidence from randomized controlled trials (as opposed to mere data), and counsel inaction until a certain ideal form of evidence— Evidence with a capital E —justifies intervening" ([25], p.42).

The scientific arguments in favour of HCQ as a means of reducing the severity of COVID-19 were of diverse types, expressed at different biological scales:

- In vitro observations where the virus in human cells is in contact with HCQ and gets destroyed [26];
- A proxy that has been observed is the level of viral load in the blood of individuals when given HCQ;
- Observation on a population with a comparison of the lethality of the disease when given HCQ vs. not given HCQ, when tested positive for Covid;
- Reports of individual experiences of doctors declaring that none of their patients had to go to the hospital when given early treatment of HCQ.

In 2025, more than 600 studies have been reported, with results ranging from neutral to positive (see https://c19hcq.org, whose meta-analysis is generally positive regarding the usefulness of HCQ in reducing hospitalisation or death due to COVID-19).

Opposing those who were considering, early in 2020, that there was enough evidence to use HCQ, others were only accepting their "gold standard" of scientific proof, RCT.

There exist many critics of RCT regarding its demonstrative value: for example, they are often multicentric (made in different places in parallel, which provokes problems of statistical aggregations with some counterintuitive results) and led mainly by the industry because of their high costs, which implies a possible bias in the expression of results. For a relatively complete review of critics, see [32], which also explains that biological plausibility needs to be established before leading a RCT.

One can consider the call for an RCT a category mistake (although it was widely reported in the media). RCT has been used in one domain, without much contestation for a long time: to allow a new drug to be marketed, one had to demonstrate the superiority of the new drug over existing ones in a disease [12]. However, HCQ has been used since the Second World War, and it was available without a prescription in France until 2019; thus, the absolute requirement for an RCT to use HCQ is not a medical necessity [8]. This is why we refer to it as a bureaucratic argument: it rests on bureaucratic logic, not medical logic.

4.3 Which side are you on? Ad hominem attacks and the association fallacy

It is worth mentioning that, although *ad hominem* arguments are usually considered a fallacy, they are nevertheless acceptable in some contexts, for example, when used to attack an unreliable witness in a legal court.

What is interesting in the case of Raoult is that the *ad hominem* attacks seem to follow an independent dynamic, with a cumulative expression in the media. Several types of arguments about Raoult, from different categories and sometimes defeated, were then used over time:

- Raoult was accused of being visibly dishonest because he was publishing too much (this argument is an exact inversion of the usual proof for quality in research: here, a h-index of 117 is a sign of bad research, whereas it is used as a positive element in an academic evaluation) [47]. The explanation for this view is that publishing too much is a sign that one prioritises one's career over knowledge (which remains unproven). In the NewYork Times, an article with a flashy title and a rather scary picture of Raoult was summarizing the downfall of the scientist [52].
- People outside his field mistreated him: for example, someone running an economics and finance blog said he was a liar [4] because he proposed his results and interpretations rather than following the RCT method. His knowledge, as inferred from his very high citation index, as a lifelong specialist in infectious diseases, had thus no value: even if HCQ was recognised as very common in respiratory diseases due to Coronaviruses [51].
- He was accused of bad behavior against colleagues, which was identified in particular thanks to some anonymous attacks, which can be heard in a documentary in 2022 [48], but was already signaled before the crisis and was repeatedly referred to during the crisis [16].
- There was an accusation of being involved in financial cheating. However, a deeper analysis of the case revealed that this issue concerned the institute's president, not the scientific crew [13].
- Even his family conflicts were exposed in the newspapers, with a daughter complaining about her father [14].

Public opinion on social networks, journalists, and TV 'experts' took positions for or against Raoult based on the methodological reference described above. A poll conducted for LCI in May 2020 showed that 45% of French people trusted Raoult, while 35% held a negative opinion of him.[5] Only 20% of participants stated they

[5]As observed in [11], at that time his institute in Marseille was one of the very few places in France that massively tested the population and responded to a frightened public. In the rest of France, people were advised not to go to the hospital unless they could no longer breathe autonomously.

had no opinion. The poll also showed that extreme political parties, both on the right and on the left, supported Raoult.

This association is curious: why would party preferences be related to one's drug choice? The heated discussion surrounding Raoult led to people removing friends from their Facebook pages. In March 2020, Trump and later Bolsonaro showed confidence in Raoult's treatment (see Figure 2 for a visual reconstruction of the main arguments circulated in March-May 2020 around the HCQ treatment). Some arguments *by association* began to appear: HCQ is associated with the right wing, so left-wing people reject it (similar to ivermectin). During this period, molecules had political opinions! A similar *association fallacy (or guilt by association)* [56] may explain the attacks on those who defended Raoult in the RCT: they were perceived as easily influenced and quickly assimilated into Raoult's camp. If you agree with Raoult on the RCT topic, you agree with him on everything. *Tertium non datur*: either you are for or against.

The association fallacy is a type of *ad hominem* argument with the following form (or schema) [55]:

- X makes the same claim as Y on a certain topic;
- Y tells atrocious and horrible nonsense on another topic;
- Then, X is surely dubious.

4.4 Dynamics of argumentation

In this section, we highlight features identified in the two debates studied in this paper. Some of these features are (to the best of our knowledge) characteristics that have not been described before in the literature.

4.4.1 Sliding arguments

Possibly the most salient feature that we identified in the two cases is what we call *sliding arguments*. When arguments are chained, the topic may change with each answer to an argument. In the Hot Coffee case, this occurred when McDonald's responsibility for serving a dangerously hot coffee shifted to a discussion on tort reform. In the HCQ case, the answer to an observational paper about HCQ was to ignore the original point and redirect the discussion to what science is or should be, and then to Raoult's supposedly bad behavior. These shifting arguments illustrate a departure from the central issue of the discussion. This way of never trying to see a discussion through its end and switching topics is the informal fallacy of *irrelevant conclusion*: "the informal fallacy of presenting an argument whose conclusion fails to address the issue in question" [57].

4.4.2 Clones

Also close to the fallacy of irrelevant conclusion is the fact that the mere repetition of a (potentially false) claim increases its familiarity to the hearer, and thus contributes to making the claim stronger (cf. the validity effect in Section 2.1).[6] It is worth mentioning that repetition is not considered in formal argumentation models. Because of the pivotal role played by such a feature, we believe that the formalization of real debates should take into account argument clones and their impact on the debate and assessment by individuals.

4.4.3 Shifting proof standards

In the literature on formal argumentation, individuals are usually ssumed to share the same proof standard. However, in the dynamics of real discussions, it can change (sometimes without explicit acknowledgement). In the Hot Coffee case, the shift was evident when the debate reached the tribunal. As Toulmin observed, the standards of argument relevant in a justice court are field-dependent. The shift that occurred in the HCQ case is more subtle. At the beginning of the discussion on HCQ, RCT was required to prove that HCQ works (*argument by established rule*, see Figure 2). However, when The Lancet paper claimed that HCQ is dangerous, it was immediately cited by journalists and ministers. Yet, the (later retracted) paper was based on a meta-analysis, which is, of course, extremely different from an RCT.

4.4.4 Zombie information

Sometimes, participants in a debate do not see certain arguments (and even facts), especially when they support the opposing view. In the Hot Coffee documentary [49] we see several people who, years after the case was concluded, still had no idea of how severe the burns suffered by Liebeck were. They recalled the case and still found it ridiculous that the lady could walk out with a lot of money. Yet, when they were shown the pictures, they were horrified and regretted their thoughts.

In the HCQ debate, when The Lancet's paper was retracted, this information seemed not visible to a portion of the public, as represented by the grey rectangle in

[6]This did not escape to Lewis Carroll, cf. *The Hunting of the Snark* poem: "Just the place for a Snark!" the Bellman cried [...] Just the place for a Snark! I have said it twice: That alone should encourage the crew. Just the place for a Snark! I have said it thrice: What I tell you three times is true." [...] The proof is complete, If only I've stated it thrice."

Figure 3. In [23], the authors investigate the phenomenon whereby retracted papers continue to be cited. They conclude that this is mainly because individuals share the retraction news only with those they expect to know the publication in the first place. This may explain why some wrong opinions are hard to dislodge.

4.4.5 The *hedgehog argument*

As discussed earlier, constructing a scientific theory requires time, which inevitably brings uncertainty and the willingness to revise one's stance. Scientific discussions can then be represented as argumentative chains with no (virtual) limit. In the HCQ debate, on the other hand, we observed an argument type we call *hedgehog*, which starts from a central claim and presents a sequence of counterarguments, with each new counterargument appearing as soon as the previous one has been addressed. In Figure 5, for instance, the central claim (the efficacy of HCQ) is attacked by one counterargument stating that the evidence offered by Raoult's group is insufficient, to which Raoult responded, claiming that the priority was to save human lives. At that point, another counterargument is presented, about the fact that only RCT is a valid type of proof, to which it is responded that RCT presents critical ethical concerns, and so on. So, whereas a scientific discussion can chain many arguments, each responding to a previous one, the length of media argument chains appears to be limited to 2 (cf. Figure 6).

The oversimplification of facts in the media was also a salient characteristic in the hot coffee debate (see Figure 1). Liebeck was depicted as an old lady drinking coffee (which is obviously hot) while driving a car and earning millions from her negligence at the expense of McDonald's. The richness of the legal debate, the 'details' of the facts (the coffee was not just normally hot, it was dangerously hot as testified by a surgeon during the trial, the fact that Liebeck was not carelessly driving a car, but she was in the passenger's seat in a vehicle parked, the information that McDonald had received more than 700 complains due to hot coffee burns etc.) were not given. No images of the burns were shown to the public in the media. In Saladoff's documentary, when these photos are displayed, viewers are shocked and begin to revise their opinions of this famous case. As in science, legal debates are rich and long. Prakken, for example, is a proponent of viewing argumentation theory as a foundational model for legal reasoning. In this vein, he developed dialogue systems to formalize how legal arguments can be presented, attacked, and defended in a back-and-forth style. Like science, law is not purely deductive but involves defeasible reasoning, where conclusions can be revised or overturned in light of new arguments or evidence [42].

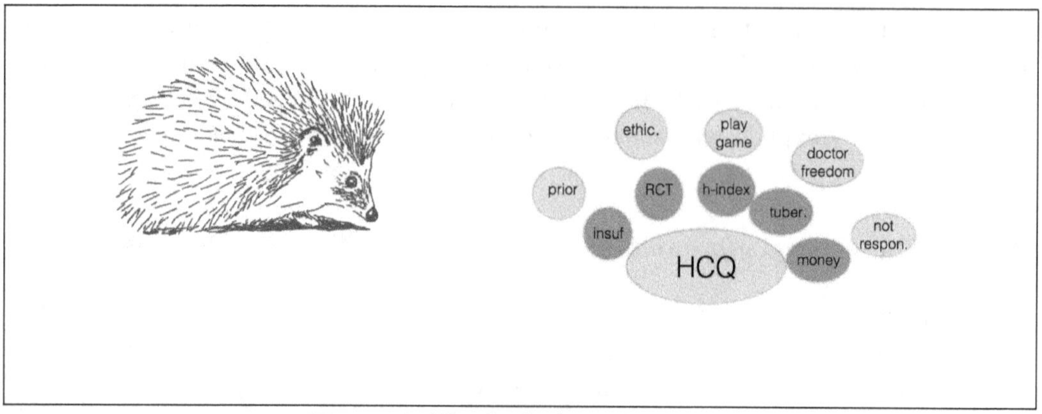

Figure 5: The hedgehog argument.

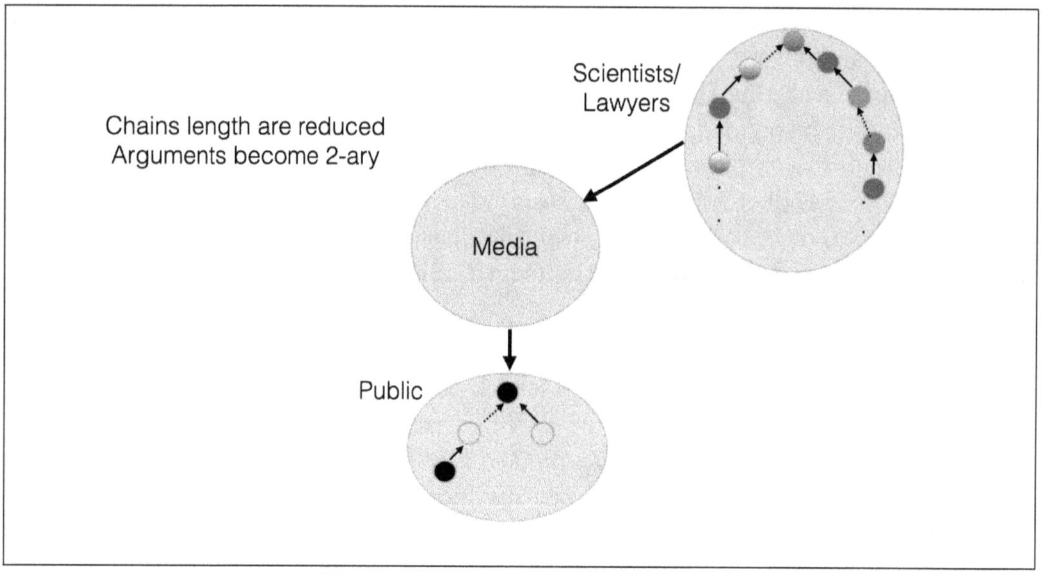

Figure 6: Argumentative chains.

5 Conclusion

In this paper, we have analysed two cases: the U.S. legal case Liebeck v. McDonald's Restaurants and the HCQ debate about the efficacy of HCQ as a treatment for COVID-19.

These two examples illustrate a similar distortion by mainstream media, employing argumentative methods that are not typically associated with the world observed by journalists, thereby imposing their own rationality.

We highlighted several indicators of anomaly (some, such as the hedgehog argument, are novel in the literature), including the multiplicity and repetition of similar arguments on one side of the problem, the lack of information from the opposite side, sliding argumentation, zombie arguments, etc.

The argumentation spaces and their interactions in the hot coffee and HCQ debates can be represented as in Figures 1 and 7, respectively. Arguments and facts are exchanged and move between spaces, each space adhering to its own standards of proof.

In both cases, the media reframed the claims made by the other fields and simplified the selected information. Figure 6 focuses on the effects of this simplification on the structure of the debate. Long (legal or scientific) argumentation exchanges of fine arguments (represented by different grades of grey) are truncated in their length (cf. hedgehog argument), and grey arguments are defuzzified and transformed into crisp black or white ones. A coffee so hot that it can cause severe burns becomes a trivial hot coffee, and a discussion on the efficiency of a treatment becomes a miracle treatment in the HCQ debate. Thus, despite the obvious difference between these cases (a legal case in the first and a scientific dispute in the second), they display several similarities.

However, one difference worth noting is that the legal debate over hot coffee had virtually no impact on the lives of the American public. In contrast, the scientific debate over a treatment for COVID-19 had a huge impact on the lives (and deaths) of scared populations around the world and, so, on public policy decisions. This may explain the differences in the circulation of the arguments between the different spheres (private/legal/public in the Liebeck's case and scientists/public/public policy in the HCQ case): the moving of arguments back-and-forth from the scientific sphere to the public and the public policy ones were greater than what we see in the reconstruction of Liebeck legal case. In that sense, Luhmann's [27] view that the media is an important force in defining reality and can have a direct impact on policy aligns with our observations and analysis. In future work, we will model the shortcuts we have observed in both cases to exploit our preliminary models.

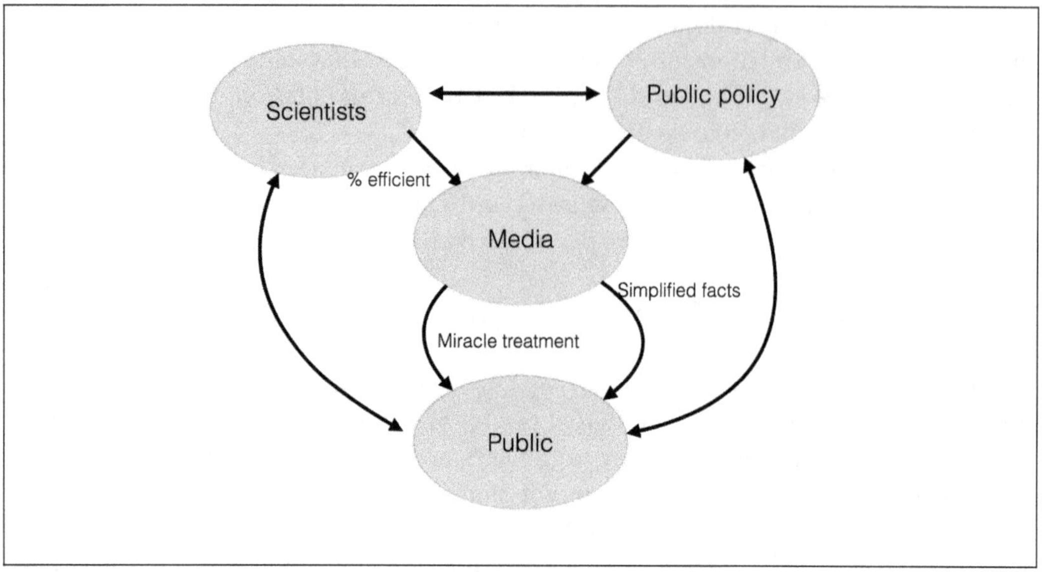

Figure 7: The HCQ argumentation spaces.

References

[1] AAVV. Consiglio di Stato in sede giurisdizionale (Sezione Terza). Ordinanza - n. 07097/2020 reg.prov.cau. url: https://tinyurl.com/2wjbhuf9, 2020.

[2] E. Anderson. Democracy, public policy, and lay assessments of scientific testimony. *Episteme*, 8(2):144–164, 2011.

[3] J. Barbot. *Les malades en mouvement. La médecine et la science à l'épreuve du sida.* Balland, 2002.

[4] O. Berruyer. Les mensonges de Didier Raoult pour promouvoir la chloroquine et faire oublier le reste. *Les Crises*, 31/03/2020, 2020. URL `https://tinyurl.com/4c54japc`.

[5] P. Bourdieu. The specificity of the scientific field and the social conditions of the progress of reason. *Social Science Information*, 14(6):19–47, 1975.

[6] P. Bourdieu. *Distinction: a Social Critique of the Judgement of Taste.* London: Routledge, 1984.

[7] P. Bourdieu. *The Field of Cultural Production.* Columbia University Press, 1993.

[8] M. Bracconi, C. Hervé, and P. Pirnay. Réflexions éthiques sur le principe de l'autonomie du patient. *Cell Research*, 23 (12), 2017. URL `https://tinyurl.com/4mckzrc6`.

[9] T. Callréus. The randomised controlled trial at the intersection of research ethics and innovation. *Pharmaceutical Medicine*, 36(5):287–293, 2022.

[10] M. Daveyand, S. Kirchgaessner, and S. Boseley. Surgisphere: governments and who changed Covid-19 policy based on suspect data from tiny us company. *The Guardian - Australia*, 2020.

[11] P. Descamps. Une médecine sous influence. *Le Monde diplomatique*, 800, November 2020.

[12] N. S. Downing, J. A. Aminawung, and N. D. Shah. Clinical trial evidence supporting fda approval of novel therapeutic agents, 2005-2012. *JAMA*, 311 (4), 2014. URL `https://sl1nk.com/FVUJQ`.

[13] FranceInfo. Marseille: l'IHU du professeur Didier Raoult perquisitionné plusieurs heures dans le cadre d'une enquête financière. *France Info*, 2021. URL `https://sl1nk.com/rt8nm`.

[14] V. Garcia. Magali Carcopino-Tusoli : "Mon père Didier Raoult se rêvait prix Nobel, il est devenu leader des complotistes. *L'express*, 36(5), 2023. URL `https://tinyurl.com/epn5mmnf`.

[15] P. Gautret, J-C. Lagier, P. Parola, V.T. Hoang, L. Meddeb, M. Mailhe, B. Doudier, J. Courjon, V. Giordanengo, V.E. Vieira, H. Tissot Dupont, S. Honoré, P. Colson, E. Chabrière, B. La Scola, J-M. Rolain, P. Brouqui, and D. Raoult. Retracted: Hydroxychloroquine and azithromycin as a treatment of COVID-19: results of an open-label non-randomized clinical trial. *International Journal of Antimicrobial Agents*, July, 2020. URL `https://pubmed.ncbi.nlm.nih.gov/32205204/`.

[16] B. Gilles. Didier Raoult accusé d'avoir pratiqué des essais non déclarés contre la tuberculose avec des complications médicales graves. *Libération*, 22 octobre 2021, 2021. URL `https://tinyurl.com/3ymbs78z`.

[17] G. T. Goodnight. The personal, technical, and public spheres of argument: a speculative inquiry into the art of public deliberation. *Journal of the American Forensic Association*, 18:214–227, 1982.

[18] G. T. Goodnight. Science and technology controversy: A rationale for inquiry. *Argumentation and Advocacy*, 42(1):26–29, 2005.

[19] E.S. Inch and K.H. Tudor. *Critical Thinking and Communication. The Use of Reason in Argument.* Pearson, 2015.

[20] S. Jackson. Studying controversies: A path for expansion of argumentation theory. *Argumentation*, 39:509–532, 2025.

[21] G. E. Kaebnick. Science doesn't work that way. *Boston Review*, April 30, 2021, 2021.

[22] T. Kuhn. *The structure of scientific revolutions.* The University of Chicago Press, 1962.

[23] T. LaCroix, A. Geil, and C. O'Connor. The dynamics of retraction in epistemic networks. *Philosophy of Science*, 88:415–438, 2021.

[24] B. Latour and S. Woolgar. *Laboratory Life : The Social Construction of Scientific Facts.* Sage Publications, 1979.

[25] M. Lipsitch. Good science is good science. In *Thinking in a Pandemic. The Crisis of Science and Policy in the Age of COVID-19.* Boston Review, Penguin Random House, 2020.

[26] J. Liu, R. Cao, M. Xu, X. Wang, H. Zhang, H. Hu, Y. Li, Z. Hu, W. Zhong, and M. Wang. Hydroxychloroquine, a less toxic derivative of chloroquine, is effective in inhibiting sars-cov-2 infection in vitro. *Cell Discovery*, 16(1), 2020.

[27] N. Luhmann. *The Reality of the Mass Media.* Stanford University Press, 2000.

[28] J. E. Mahon. The definition of lying and deception. In E. N. Zalta, editor, *The Stanford Encyclopedia of Philosophy.* Metaphysics Research Lab, Stanford University, 2016. URL `https://tinyurl.com/3s2n8ty9`.

[29] M. R. Mehraa, F. Ruschitzka, and A. N. Patel. Retraction - Hydroxychloroquine or chloroquine with or without a macrolide for treatment of COVID-19: a multinational registry analysis. *The Lancet*, 395, 2020.

[30] H. Mercier and D. Sperber. *The Enigma of Reason.* Harvard University Press, 2017.

[31] D. Mielke and V. Rohde. Randomized controlled trials – a critical re-appraisal. *Neurosurgical Review*, 2020.

[32] B. D. Nimavat, K. G. Zirpe, and S. K. Gurav. Retracted: Hydroxychloroquine and azithromycin as a treatment of COVID-19: results of an open-label non-randomized clinical trial. *Indian Journal of Critical Care Medicine*, 56:215–222, 2020.

[33] C. O'Connor. Retraction and information zombies. In *CPNAS Wonder Workshop: Knowledge Commons and the Future of Democracy. Online through Cultural Programs of the National Academy of Sciences*, 2023.

[34] J.O. Ovosi, M.S. Ibrahim, and B.O. Bello-Ovosi. Randomized controlled trials: Ethical and scientific issues in the choice of placebo or active control. *Annals of African Medicine*, 16(3):97–100, 2017.

[35] M. Pera. Rhetoric and scientific controversies. In M. Pera P. Machamer and A. Baltas, editors, *Scientific Controversies: Philosophical and Historical Perspectives*, pages 50–66. Oxford University Press, 2000.

[36] C. Perelman and L. Olbrechts-Tyteca. *La nouvelle rhétorique: Traité de l'argumentation.* Paris: Presses universitaires de France, 1958.

[37] C. Perelman and L. Olbrechts-Tyteca. *The New Rhetoric: A Treatise on Argumentation.* otre Dame, IN: University of Notre Dame Press, 1969.

[38] P. Pignarre. Ce que masque l'affaire Raoult ? L'infantilisation des patients. *Nouvel Observateur*, 19 septembre 2020, 2020. URL `https://tinyurl.com/bhdy962j`.

[39] S. Polberg and A. Hunter. Empirical evaluation of abstract argumentation: Supporting the need for bipolar and probabilistic approaches. *International Journal of Approximate Reasoning*, 93:487–543., 2018.

[40] K. Popper. *The Open Society and its Enemies.* London: Routledge, 1945.

[41] A. Pradelle, S. Mainbourg, S. Provencher, E. Massy, G. Grenet, and J-C Lega. Deaths induced by compassionate use of hydroxychloroquine during the first COVID-19 wave: an estimate. *Biomedicine and Pharmacotherapy*, 171, 2024. URL `https://pubmed.ncbi.nlm.nih.gov/38171239/`.

[42] H. Prakken. A formal model of adjudication dialogues. *Artificial Intelligence and Law*, 16(4):305–328, 2008.

[43] W.V. Quine. *Word and Object.* Cambridge, Mass., MIT Press, 1960.

[44] D. Raoult. *Commission d'enquête COVID-19 : Audition de Didier Raoult, 24 Juin 2020, réponse à Philippe Berta, Député MODEM*, 2020. URL `https://tinyurl.com/`

4ast2vnh. Accessed on July 14, 2024.

[45] C. Hackett Renner. Validity effect. In R. F. Pohl, editor, *Cognitive Illusions. A Handbook on Fallacies and Biases in Thinking, Judgement and Memory*, pages 201–213. Taylor & Francis, 2004.

[46] RetroReport. *The Misunderstood McDonald's Hot Coffee Lawsuit*, 2019. URL https://tinyurl.com/4d4mzdha. Accessed on July 14, 2024.

[47] M. Roche. Didier Raoult a-t-il vraiment signé «3500 publications internationales» dans sa carrière ? *Libération Check News*, 22/09/2020, 2020. URL https://tinyurl.com/msjsvchv.

[48] G. Rof. Didier Raoult, le savant flou, France 2: la dérive d'un système incontrôlé et incontrôlable. *Le Monde*, 16/11/22, 2022. URL https://tinyurl.com/53uprcas.

[49] S. Saladoff. *Hot Coffee*, 2011.

[50] N. Salliou and R.N Thomopoulos. Modelling multicriteria argument networks about reduced meat consumption. In *FOODSIM 2018-10th International Conference on Simulation and Modelling in the Food and Bio-industry*, pages 1–6, 2018.

[51] A. Savarino, L. di Trani, I. Donatelli, R. Cauda, and A. Cassone. New insights into the antiviral effects of chloroquine. *The Lancet Infectious disease*, 6(2), 2006. URL https://tinyurl.com/5cpzuz3m.

[52] S. Sayare. He was a science star. then he promoted a questionable cure for Covid-19. *NewYork Times*, 12/05/20, 2020. URL https://tinyurl.com/475num2r.

[53] I. Stengers. Ce que nous apprend la pandémie. Pour un atterrissage des sciences. *Esprit*, 2021(3):37–47, 2021.

[54] S.E. Toulmin. *The uses of argument.* Cambridge: Cambridge University Press, 1958.

[55] Douglas Walton. *Ad Hominem Arguments.* Tuscaloosa: University of Alabama Press, 1998.

[56] Wikipedia. *Association fallacy.* URL https://en.wikipedia.org/wiki/Association_fallacy. Accessed on December 7, 2025.

[57] Wikipedia. *Irrelevant conclusion.* URL https://en.wikipedia.org/wiki/Irrelevant_conclusion. Accessed on December 7, 2025.

[58] L. Yu and L. van der Torre. The A-BDI metamodel for human-level AI. In T. Agotnes and D. Doder, editors, *Logic and Argumentation. CLAR 2025. Lecture Notes in Computer Science*, volume 15712, pages 1–2. Springer, Singapore, 2025.

[59] L. Yu, L. van der Torre, and R. Markovich. Thirteen challenges in formal and computational argumentation. In D. Gabbay, G. Kern-Isberner, G.R. Simari, and M. Thimm, editors, *Handbook of Formal Argumentation, Volume 3*, pages 890–976. College Publications, 2024.

Received July 2025

On the Structure of Dual-line Standard Contradictions and their General Forms in First-order Logic

Xingxing He, Jia Xu
School of Mathematics, Southwest Jiaotong University, Chengdu, 610031, China

Yingfang Li*
School of Computing and Artificial Intelligence, Southwestern University of Finance and Economics, Chengdu 611130, China
liyf@swufe.edu.cn

Jun Liu
School of Computing, Ulster University, Belfast BT15 1ED, Northern Ireland, UK

Abstract

Contradiction separation (CS) and its first-order version S-CS are multi-clause inference schemes for clausal refutation. They isolate a standard contradiction core within a clause set and derive a propagated clause from the remaining literals. This paper develops structural characterizations of non-unit standard contradictions that make such cores explicit and easier to identify. In propositional logic, we introduce a canonical dual-line family and prove that every instance is a standard contradiction. We study admissible literal extensions, define ladder structures as maximal dual-line extensions, and present a regular triple-line family with constructive generation schemes. We also analyze how dual-line cores compose via clause connections and give sufficient conditions under which the composed clause set remains a standard contradiction. In first-order logic, we exhibit clause families that are not standard contradictions syntactically but become standard contradictions after suitable instantiation and controlled clause reuse.

*The corresponding author.

1 Introduction

Automated reasoning (AR) studies algorithmic proof search in formal calculi [1, 11]. In clausal theorem proving, a conjecture φ is typically handled by clausifying its negation $\neg\varphi$ and showing that the resulting clause set is unsatisfiable, usually by deriving the empty clause in a sound refutational calculus such as resolution and its extensions [12]. For a new inference rule to be relevant to AR, soundness is only the starting point. Its applicability must be detectable during proof search, and the clauses it derives must interact well with the prover's standard mechanisms for simplification, subsumption, and redundancy control [9, 10].

Contradiction separation (CS) and its first-order version S-CS were introduced in this setting [19] and have since been studied further in work on multi-clause deduction [2, 18, 20]. Unlike binary resolution or hyper-resolution, which have fixed arity [4], CS/S-CS may use several clauses and literals in one step. Each participating clause C_i is decomposed into a contradictory part C_i^- and a residual part C_i^+. If the conjunction $\bigwedge_i C_i^-$ is a standard contradiction, then the disjunction $\bigvee_i C_i^+$ is derived as a new clause. From the viewpoint of theorem proving, this is a macro-step: a contradiction distributed across several clauses is condensed into a derived lemma that can be fed back into the ordinary saturation loop. In practice (e.g., in CSE and its integration with E [13] as CSE-E [3, 21]), such derived clauses are injected into the prover's usual search loop and handled by the same simplification and redundancy mechanisms as ordinary clauses [14, 15]. The role of standard contradictions in AR is therefore operational as well as logical: they determine when a genuinely multi-clause inference can be triggered.

The main difficulty is that the definition of a standard contradiction is global and combinatorial [19]. It quantifies over all choices of one literal from each clause. A direct applicability test based on the definition is therefore too expensive to be attractive inside proof search. What CS/S-CS needs instead is structural information [5, 6, 16]: recognizable clause patterns that expose contradiction cores without repeated brute-force checking. This makes the identification of standard-contradiction cores the central issue for bringing CS/S-CS closer to practical use in AR.

The present paper studies that issue for non-unit clause sets. In propositional logic, we isolate canonical families of standard contradictions, examine how they behave under literal extension and clause-set composition, and show how larger cores can be built from smaller ones. In first-order logic, we identify clause families whose standard-contradiction property is not visible at the syntactic level but emerges after suitable instantiation and controlled clause reuse. The contribution is thus proof-theoretic rather than implementational: we do not present an experimental evaluation, but we do give explicit structures and applicability conditions that can support future core-detection heuristics and future integrations of CS/S-CS into prover search.

Contributions. The paper makes three main contributions.

1. We introduce a canonical dual-line family and prove that every instance is a standard contradiction. We then study admissible literal extensions, define ladder structures as maximal dual-line extensions, and analyze when dual-line cores are preserved under clause connections. These results provide explicit templates for recognizing non-unit contradiction cores and for understanding when such cores survive extension or composition.

2. We present a regular triple-line family, prove that it is a standard contradiction, and give constructive generation schemes. This enlarges the class of recognizable core patterns beyond the minimal dual-line case and shows that CS applicability is supported by a broader range of structured clause configurations.

3. We identify representative first-order clause families that are not standard contradictions syntactically but become standard contradictions after suitable instantiation and controlled clause reuse. The results cover both identical and distinct predicate symbols, and terms built from constants, variables, and function symbols. They clarify that, for S-CS, applicability depends not only on clause shape but also on how substitutions expose complementary structure across several clauses.

2 Preliminaries

We fix basic syntax and semantics for propositional and first-order clause reasoning [5, 7, 8, 17, 19].

Let Var be a set of propositional variables. A literal is p or $\neg p$ ($p \in \mathsf{Var}$), and $p, \neg p$ are complementary. A clause is a finite disjunction $C = L_1 \vee \cdots \vee L_n$ ($n \geq 1$); a unit clause has $n = 1$; the empty clause is $\Box$. A clause set (CNF) is a finite set $S = \{C_1, \ldots, C_m\}$, read as $C_1 \wedge \cdots \wedge C_m$. A valuation $v : \mathsf{Var} \to \{0, 1\}$ extends to literals by $v(\neg p) = 1 - v(p)$. Satisfaction is defined by $v \models C$ iff $v(L_i) = 1$ for some i, and $v \models S$ iff $v \models C$ for all $C \in S$. S is satisfiable if some $v \models S$, and unsatisfiable otherwise.

A first-order language $\mathcal{L}$ consists of constants, function symbols, predicate symbols (and equality if present). Terms are built from variables, constants, and functions. Atomic formulas are $P(t_1, \ldots, t_n)$ (and $t \approx s$ if equality is included). First-order literals, clauses, and clause sets are defined as in the propositional case.

An $\mathcal{L}$-structure $\mathcal{M}$ has a nonempty domain D and interprets symbols over D; a variable assignment σ maps variables to D. Truth of literals is evaluated under $(\mathcal{M}, \sigma)$ in the standard way. Define $(\mathcal{M}, \sigma) \models C$ iff $(\mathcal{M}, \sigma) \models L$ for some $L \in C$, and $(\mathcal{M}, \sigma) \models S$ iff $(\mathcal{M}, \sigma) \models C$ for all $C \in S$. Write $\mathcal{M} \models S$ if $(\mathcal{M}, \sigma) \models S$ for all σ. S is satisfiable if some $\mathcal{M} \models S$, and unsatisfiable otherwise.

A substitution is a finite mapping $\theta = \{t_1/x_1, \ldots, t_k/x_k\}$ from pairwise distinct variables to terms. Application to terms/literals/clauses/clause sets is homomorphic and denoted by $(\cdot)\theta$; $E\theta$ is an instance of E.

Definition 1. *[19] Let $S = \{C_1, C_2, \ldots, C_m\}$ be a clause set. If for any $(p_1, \ldots, p_m) \in \prod_{i=1}^{m} C_i$, there exists at least one complementary pair among $\{p_1, \ldots, p_m\}$, then $S = \bigwedge_{i=1}^{m} C_i$ is called a standard contradiction. If $\bigwedge_{i=1}^{m} C_i$ is unsatisfiable, then $S = \bigwedge_{i=1}^{m} C_i$ is called a quasi-contradiction.*

Note that these notions depend only on the underlying set of clauses and literals: permuting clauses or reordering literals within clauses does not affect whether S is a standard contradiction or a quasi-contradiction. We now recall the inference rule that extracts a new clause by separating a standard contradiction from a given clause set.

Definition 2. *[19] (Contradiction Separation Rule in Propositional Logic) Assume a clause set $S = \{C_1, C_2, \ldots, C_m\}$. The following inference rule that produces a new clause from S is called a contradiction separation rule, in short, a CS rule:*

For each C_i $(i = 1, \ldots, m)$, separate it into two sub-clauses C_i^- and C_i^+ such that

1. *$C_i = C_i^- \vee C_i^+$, where C_i^- and C_i^+ have no common literals;*
2. *C_i^+ can be an empty clause itself, but C_i^- cannot be an empty clause;*
3. *$\bigwedge_{i=1}^{m} C_i^-$ is a standard contradiction.*

The resulting clause $\bigvee_{i=1}^{m} C_i^+$, denoted as $\mathcal{C}_m(C_1, C_2, \ldots, C_m)$, is called a contradiction separation clause (CSC) of $C_1, C_2, \ldots, C_m$, and $\bigwedge_{i=1}^{m} C_i^-$ is called a separated contradiction (SC).

Definition 3. *[19](Standard Contradiction Separation Rule in First-Order Logic) Suppose a clause set $S = \{C_1, C_2, \ldots, C_m\}$ in first-order logic. Without loss of generality, assume that there do not exist the same variables among $C_1, C_2, \ldots, C_m$ (If there exist the same variables, there exists a renaming substitution that makes them different). The following inference rule that produces a new clause from S is called a standard contradiction separation rule, in short, an S-CS rule:*

For each C_i $(i = 1, 2, \ldots, m)$, firstly apply a substitution σ_i to C_i (σ_i could be an empty substitution, but not necessarily the most general unifier), denoted as $C_i^{\sigma_i}$; then separate $C_i^{\sigma_i}$ into two sub-clauses $C_i^{\sigma_i -}$ and $C_i^{\sigma_i +}$ such that

1. *$C_i^{\sigma_i} = C_i^{\sigma_i -} \vee C_i^{\sigma_i +}$, where $C_i^{\sigma_i -}$ and $C_i^{\sigma_i +}$ have no common literals;*
2. *$C_i^{\sigma_i +}$ can be an empty clause itself, but $C_i^{\sigma_i -}$ cannot be an empty clause;*
3. *$\bigwedge_{i=1}^{m} C_i^{\sigma_i -}$ is a standard contradiction, that is for any $(x_1, \ldots, x_m) \in \prod_{i=1}^{m} C_i^{\sigma_i -}$, there exists at least one complementary pair among $\{x_1, \ldots, x_m\}$.*

The resulting clause $\bigvee_{i=1}^{m} C_i^{\sigma_i +}$, denoted as $C_m^{s\sigma}(C_1, \ldots, C_m)$ (here "s" means "standard", $\sigma = \cup_{i=1}^{m} \sigma_i$, σ_i is a substitution to C_i, $i = 1, \ldots, m$), is called a standard contradiction separation clause (S-CSC) of $C_1, \ldots, C_m$, and $\bigwedge_{i=1}^{m} C_i^{\sigma_i -}$ is called a separated standard contradiction (S-SC).

Definition 4. *[19] For a standard contradiction S, if deleting clauses results in a clause set S′ that is still a standard contradiction, the deleted clauses are called redundant clauses.*

We use standard contradiction and quasi-contradiction as in Definition 1. Throughout the paper, a standard-contradiction core is any (possibly instantiated) subcollection of clauses that forms a standard contradiction and serves as the contradictory part in a CS/S-CS application. In particular, in Definition 2 the separated conjunction $\bigwedge_{i=1}^{m} C_i^{-}$ is a standard-contradiction core (called an SC in the original terminology), and in the first-order case $\bigwedge_{i=1}^{m} C_i^{\sigma_i -}$ plays the same role (an S-SC). We refer to the task of locating such cores in a clause set or in a refutation trace as standard-contradiction core identification (or core identification when the context is clear).

In what follows, we restrict attention to clause sets in which every literal occurring in the set has its complementary literal occurring somewhere in the same set. This excludes degenerate situations (e.g., literals that can never participate in a complementary pair) and keeps the structural analysis focused on the contradiction-core patterns relevant to CS/S-CS.

3 Structures of non-unit standard contradictions in propositional logic

3.1 Dual-line and triple-line standard contradictions

We start with dual-line standard contradictions, which provide the simplest non-unit template that guarantees a standard contradiction. Although unit clauses are excluded, the dual-line pattern already captures a genuinely multi-clause configuration that can serve as a standard-contradiction core in CS/S-CS derivations.

Moreover, dual-line cores are stable under natural construction operations. We therefore analyze how the standard-contradiction property behaves under literal extension and clause composition, and use these observations to motivate the richer families introduced later (triple-line and ladder structures). A precise characterization of the dual-line case thus provides the baseline for the subsequent definitions and results.

Definition 5. *Let $S = \{C_1, C_2, \cdots, C_{m+1}\}$ $(m \geq 3)$ be a clause set with a dual-line structure, where $C_1 = p_1 \vee p_2$, $C_i = \neg p_{i-1} \vee p_i$ $(i = 2, \cdots, m)$, and $C_{m+1} = \neg p_{m-1} \vee \neg p_m$.*

For a clause set $S = \{C_1, C_2, \cdots, C_{m+1}\}$ with a dual-line structure, C_1 contains literals p_1 and p_2, the positive literal p_i in C_i $(1 \leq i \leq m)$ has its complement $\neg p_i$ in C_{i+1}, and C_{m+1} contains literals $\neg p_{m-1}$ and $\neg p_m$ as shown in Table 1.

C_1	C_2	C_3	C_4	$\cdots$	$\cdots$	C_{m-1}	C_m	C_{m+1}
							p_m	$\neg p_m$
						p_{m-1}	$\neg p_{m-1}$	$\neg p_{m-1}$
					$\cdots$	$\neg p_{m-2}$		
				$\cdots$	$\cdots$			
			p_4	$\cdots$				
		p_3	$\neg p_3$					
p_2	p_2	$\neg p_2$						
p_1	$\neg p_1$							

Table 1: A clause set with dual-line structure

Theorem 1. *If S is a clause set with dual-line structure, then S is a standard contradiction.*

Proof. Select one literal from each C_i $(i = 1, 2, \cdots, m+1)$.

(1) If p_1 is selected from C_1 and $\neg p_1$ from C_2, then there exists a complementary pair $(p_1, \neg p_1)$.

(2) If p_1 is not selected from C_1, at least p_2 is selected from clauses C_1 or C_2. If $\neg p_{i-1}$ is selected from C_i $(i = 3, \cdots, m)$, then there exists a complementary pair $(p_j, \neg p_j)$ $(2 \leq j \leq i)$.

(3) If p_1 is not selected from C_1, at least p_2 is selected from clauses C_1 or C_2. If none of $\neg p_{i-1}$ is selected from C_i $(i = 3, \cdots, m)$, the selected literals form the set $\{p_1, p_2, \cdots, p_m\}$ or $\{p_2, p_2, \cdots, p_m\}$. Since C_{m+1} contains literals $\neg p_{m-1}$ and $\neg p_m$, there exists a complementary pair $(p_{m-1}, \neg p_{m-1})$ or $(p_m, \neg p_m)$.

□

A clause set satisfying the dual-line structure will be called a dual-line standard contradiction.

Next, we give a concrete example to illustrate how dual-line standard contradictions can serve as structured cores in CS-based dynamic deduction.

Example 1. *Let a clause set $S = \{C_1, \ldots, C_{15}\}$ in propositional logic, where $C_1 = p_1 \vee p_2 \vee p_8 \vee p_9$, $C_2 = \neg p_1 \vee p_2 \vee p_8 \vee p_{10}$, $C_3 = \neg p_2 \vee p_3 \vee p_9 \vee p_{11}$, $C_4 = \neg p_3 \vee p_4 \vee p_{10} \vee p_{12}$, $C_5 = \neg p_4 \vee p_5 \vee p_{11} \vee p_{13}$, $C_6 = \neg p_4 \vee \neg p_5 \vee p_{12} \vee p_{13}$, $C_7 = \neg p_8 \vee p_{14}$, $C_8 = \neg p_9 \vee p_{14}$, $C_9 = \neg p_{10} \vee p_{14}$, $C_{10} = \neg p_{11} \vee p_{14}$, $C_{11} = \neg p_{12} \vee p_{14}$, $C_{12} = \neg p_{13} \vee p_{14}$, $C_{13} = \neg p_{14} \vee p_{15}$, $C_{14} = p_{14} \vee \neg p_{15}$, $C_{15} = \neg p_{14} \vee \neg p_{15}$.*

Separated from $C_1 \sim C_6$: $C_1^- = p_1 \vee p_2$, $C_1^+ = p_8 \vee p_9$; $C_2^- = \neg p_1 \vee p_2$, $C_2^+ = p_8 \vee p_{10}$; $C_3^- = \neg p_2 \vee p_3$, $C_3^+ = p_9 \vee p_{11}$; $C_4^- = \neg p_3 \vee p_4$, $C_4^+ = p_{10} \vee p_{12}$; $C_5^- = \neg p_4 \vee p_5$, $C_5^+ = p_{11} \vee p_{13}$; $C_6^- = \neg p_4 \vee \neg p_5$, $C_6^+ = p_{12} \vee p_{13}$.

1. *According to Theorem 1, $\bigwedge_{i=1}^{6} C_i^-$ is a dual-line standard contradiction. Using the CS rule for the clauses $C_1, C_2, C_3, C_4, C_5, C_6$, we obtain a CSC involving 6 clauses: $C_{16} = \mathcal{C}_6(C_1, C_2, C_3, C_4, C_5, C_6) = p_8 \vee p_9 \vee p_{10} \vee p_{11} \vee p_{12} \vee p_{13}$.*

2. *The second standard contradiction is $(p_8 \vee p_9 \vee p_{10} \vee p_{11} \vee p_{12} \vee p_{13}) \wedge (\neg p_8) \wedge (\neg p_9) \wedge (\neg p_{10}) \wedge (\neg p_{11}) \wedge (\neg p_{12}) \wedge (\neg p_{13})$. Using the CS rule for the clauses C_{16}, C_7, C_8, C_9, C_{10}, C_{11}, C_{12}, then $C_{17} = \mathcal{C}_7(C_{16}, C_7, C_8, C_9, C_{10}, C_{11}, C_{12}) = p_{14}$.*

3. *The third standard contradiction is $(\neg p_{14} \vee p_{15}) \wedge (p_{14} \vee \neg p_{15}) \wedge (\neg p_{14} \vee \neg p_{15}) \wedge p_{14}$. Using the CS rule for the clauses $C_{13}, C_{14}, C_{15}, C_{17}$, then $C_{18} = \mathcal{C}_4(C_{13}, C_{14}, C_{15}, C_{17}) = \square$.*

The following shows binary resolution in multiple steps.
$C_{16} = \mathcal{R}(C_1, C_2) = p_2 \vee p_8 \vee p_9 \vee p_{10}$
$C_{17} = \mathcal{R}(C_{16}, C_3) = p_3 \vee p_8 \vee p_9 \vee p_{10} \vee p_{11}$
$C_{18} = \mathcal{R}(C_{17}, C_4) = p_4 \vee p_8 \vee p_9 \vee p_{10} \vee p_{11} \vee p_{12}$
$C_{19} = \mathcal{R}(C_{18}, C_5) = p_5 \vee p_8 \vee p_9 \vee p_{10} \vee p_{11} \vee p_{12} \vee p_{13}$
$C_{20} = \mathcal{R}(C_{18}, C_6) = \neg p_5 \vee p_8 \vee p_9 \vee p_{10} \vee p_{11} \vee p_{12} \vee p_{13}$
$C_{21} = \mathcal{R}(C_{19}, C_{20}) = p_8 \vee p_9 \vee p_{10} \vee p_{11} \vee p_{12} \vee p_{13}$
$C_{22} = \mathcal{R}(C_{21}, C_7) = p_9 \vee p_{10} \vee p_{11} \vee p_{12} \vee p_{13} \vee p_{14}$

$C_{23} = \mathcal{R}(C_{22}, C_8) = p_{10} \vee p_{11} \vee p_{12} \vee p_{13} \vee p_{14}$
$C_{24} = \mathcal{R}(C_{23}, C_9) = p_{11} \vee p_{12} \vee p_{13} \vee p_{14}$
$C_{25} = \mathcal{R}(C_{24}, C_{10}) = p_{12} \vee p_{13} \vee p_{14}$
$C_{26} = \mathcal{R}(C_{25}, C_{11}) = p_{13} \vee p_{14}$
$C_{27} = \mathcal{R}(C_{26}, C_{12}) = p_{14}$
$C_{28} = \mathcal{R}(C_{27}, C_{13}) = p_{15}$
$C_{29} = \mathcal{R}(C_{27}, C_{15}) = \neg p_{15}$
$C_{30} = \mathcal{R}(C_{28}, C_{29}) = \Box$

This shows that binary resolution requires 15 steps to derive $\Box$, *whereas the CS-based dynamic deduction above reaches* $\Box$ *in 3 steps.*

Remark 1. *The comparison above concerns only the length of two fixed derivations (i.e., the number of inference steps) and is intended to illustrate proof-length compression by CS as a macro-step. It should not be interpreted as a claim about runtime behaviour in modern provers, where heuristic scheduling, redundancy elimination, and indexing dominate overall performance.*

For dual-line (and later ladder) cores, certain repeated-literal extensions preserve the standard-contradiction property. In general, however, adding literals—even literals already occurring elsewhere in the clause set—need not preserve standard contradictions, and adding genuinely new literals destroys the property. We now state a sufficient condition under which specific extensions of a dual-line standard contradiction remain standard contradictions.

Proposition 1. *Let* $S = \{C_1, C_2, \cdots, C_{m+1}\}$ *be a dual-line standard contradiction. For* C_i $(i = 4, \cdots, m+1)$, *if any* $\neg p_j$ $(2 \leq j \leq i-2)$ *is added to* C_i, *and get a new clause set* S', *then* S' *is a standard contradiction.*

Proof. Select one literal from each C_i $(i = 1, 2, \cdots, m+1)$.

(1) If p_1 is selected from C_1 and $\neg p_1$ from C_2, then there exists a complementary pair $(p_1, \neg p_1)$.

(2) If p_1 is not selected from C_1, at least p_2 is selected from clauses C_1 or C_2. If $\neg p_2$ is selected from C_i $(i = 3, \cdots, m)$, then there exists a complementary pair $(p_2, \neg p_2)$.

(3) If p_1 is not selected from C_1, at least p_2 is selected from clauses C_1 or C_2. If none of $\neg p_2$ is selected from C_i $(i = 3, \cdots, m)$, select p_3 from C_3. If none of $\neg p_3$ is selected from C_i $(i = 3, \cdots, m)$, select p_4 from C_4, etc. The selected literals form the set $\{p_1, p_2, \cdots, p_m\}$ or $\{p_2, p_2, \cdots, p_m\}$. Since C_{m+1} is $\neg p_2 \vee \cdots \vee \neg p_m$, there exists a complementary pair $(p_n, \neg p_n)$ $(2 \leq n \leq m)$.

□

Remark 2. *Let $S = \{C_1, C_2, \ldots, C_{m+1}\}$ be a dual-line standard contradiction. If a literal p_0 does not occur in any clause of S, then adding p_0 to any clause destroys the standard-contradiction property. That is, for any $i \in \{1, \ldots, m+1\}$,*

$$S_1 = \{C_1, \ldots, C_{i-1}, C_i \cup \{p_0\}, C_{i+1}, \ldots, C_{m+1}\}$$

is not a standard contradiction.

Dual-line standard contradictions represent a minimal non-unit template: they guarantee a contradiction using the smallest possible positive/negative interaction pattern among the involved literals. However, many non-unit clause sets exhibit standard-contradiction cores that are not captured by a dual-line configuration and instead require a more regular multi-literal interaction.

To cover such cases, we introduce triple-line standard contradictions. In this family, the core is arranged as a uniform sequence of four-clause blocks generated from triples of literals, yielding a systematic non-unit template for more complex contradiction configurations.

Definition 6. *Let $S = \{C_1, \ldots, C_n\}$ be a clause set, where $n = 4m$ for some $m \geq 1$. We say that S has a triple-line structure generated by literals $p_1, \ldots, p_{m+1}$ (with $m \geq 2$, i.e., at least three literals) if S can be partitioned into m consecutive blocks of four clauses such that:*

- *for each block $t = 1, \ldots, m-1$, the four clauses are $C_{4t-3} = p_t \vee p_{t+1} \vee p_{t+2}, C_{4t-2} = p_t \vee \neg p_{t+1} \vee p_{t+2}, C_{4t-1} = \neg p_t \vee p_{t+1} \vee p_{t+2}, C_{4t} = \neg p_t \vee \neg p_{t+1} \vee p_{t+2}$;*
- *in the last block $t = m$, the four clauses are $C_{4m-3} = p_{m-1} \vee p_m \vee \neg p_{m+1}, C_{4m-2} = p_{m-1} \vee \neg p_m \vee \neg p_{m+1}, C_{4m-1} = \neg p_{m-1} \vee p_m \vee \neg p_{m+1}, C_{4m} = \neg p_{m-1} \vee \neg p_m \vee \neg p_{m+1}$.*

C_1	C_2	C_3	C_4	$\cdots$	C_{4m-7}	C_{4m-6}	C_{4m-5}	$C_{4(m-1)}$	C_{4m-3}	C_{4m-2}	C_{4m-1}	C_{4m}
				$\cdots$	p_m	p_m	p_m	p_m	$\neg p_m$	$\neg p_m$	$\neg p_m$	$\neg p_m$
p_3	p_3	p_3	p_3	$\cdots$	p_{m-1}	$\neg p_{m-1}$	p_{m-1}	$\neg p_{m-1}$	p_{m-1}	p_{m-1}	$\neg p_{m-1}$	$\neg p_{m-1}$
p_2	$\neg p_2$	p_2	$\neg p_2$	$\cdots$	p_{m-2}	p_{m-2}	$\neg p_{m-2}$	$\neg p_{m-2}$	p_{m-2}	$\neg p_{m-2}$	p_{m-2}	$\neg p_{m-2}$
p_1	p_1	$\neg p_1$	$\neg p_1$	$\cdots$								

Table 2: A clause set with triple-line structure

Theorem 2. *If S is a clause set with a triple-line structure, then S is a standard contradiction.*

Proof. Fix an arbitrary choice of one literal from each clause of S. Consider the $(m-1)$-st part. If the chosen literals from this part contain no occurrence of p_{m+1}, then all four clauses must have been satisfied by choices among p_{m-1} and p_m, which forces a complementary pair $(p_{m-1}, \neg p_{m-1})$ or $(p_m, \neg p_m)$ among the chosen literals. Hence, the chosen set contains p_{m+1}.

Next, consider the m-th part. If the chosen literals from this part contain no occurrence of $\neg p_{m+1}$, then the same reasoning forces a complementary pair $(p_{m-1}, \neg p_{m-1})$ or $(p_m, \neg p_m)$ among the chosen literals. Hence, the chosen set contains $\neg p_{m+1}$.

Therefore, in every selection of one literal per clause, the chosen literals contain both p_{m+1} and $\neg p_{m+1}$, or else already contain a complementary pair among p_{m-1}, p_m. In all cases, the selection contains a complementary pair $(p_i, \neg p_i)$ for some $1 \leq i \leq m+1$. Thus S is a standard contradiction. □

A clause set with a triple-line structure is called a triple-line standard contradiction.

Proposition 2. *Let* $S = \{C_1, C_2, \cdots, C_{4m}\}$ *be a standard contradiction of three lines. For any* C_i $(i = 1, \cdots, 4m)$*, add either* p_i *or* $\neg p_i$ *to* C_i*, and get a new clause set* S'*, then* S' *is a standard contradiction, although it no longer preserves the triple-line structure.*

Proof. After adding any positive literal p_i or negative literal $\neg p_i$ to C_i, When selecting literals in the j-th part $(1 \leq j < m)$, do not select p_{j+2}.

(1) Choose the added literal p_i : If $i > j$, then $\neg p_i$ will be selected in subsequent clauses. If $i < j$, then $\neg p_i$ will be selected in the preceding clauses.

(2) Choose the added literal $\neg p_i$: If $i > j$, then p_i will be selected in subsequent clauses. If $i < j$, then in the preceding clauses p_i will be selected.

Therefore, complementary pairs will still appear this way, ensuring S' remains a standard contradiction. □

Building on dual-line standard contradictions, we consider the extreme case of literal extension that still preserves the standard-contradiction property. Ladder standard contradictions formalize this setting: they are obtained by augmenting a dual-line structure with as many additional literals as possible without destroying standard contradictions. As such, they provide a convenient reference model for locating the boundary between admissible extensions and loss of the property.

Definition 7. *Let $S = \{C_1, C_2, \cdots, C_{m+1}\}$ $(m \geq 3)$, where $C_1 = p_1 \vee p_2$, $C_i = \neg p_{i-1} \vee p_i$ $(i = 2, 3)$, $C_j = \neg p_2 \vee \cdots \vee \neg p_{j-1} \vee p_j$ $(j = 4, \cdots, m)$, and $C_{m+1} = \neg p_2 \vee \cdots \vee \neg p_m$, then S is called a clause set with ladder structure.*

A clause set with a ladder structure $S = \{C_1, C_2, \cdots, C_{m+1}\}$ in Definition 7 is formed by adding the maximum number of literals to a dual-line standard contradiction.

Theorem 3. *If S is a clause set with a ladder structure, then S is a standard contradiction.*

Proof. The proof is similar to Proposition 1. □

A clause set satisfying Definition 7 is called a ladder standard contradiction. We introduce ladder structures primarily as an extremal reference point: they are maximal dual-line extensions that still satisfy the standard-contradiction property. If additional literals are appended beyond this maximal form, the property may be lost; in that case, it can be restored only by adding suitable auxiliary clauses, as discussed below.

Since ladder standard contradictions are obtained by saturating a dual-line template under literal extension, their basic structural features follow from the dual-line case, and their composition behaviour is covered by the principles in Section 4. We therefore do not develop their fundamental properties separately.

3.2 Clause addition strategy

Ladder standard contradictions are maximal with respect to literal extension: adding a further (previously existing) literal to a clause typically destroys the standard-contradiction property. To preserve the property after such an extension, one must introduce additional clauses that compensate for the new literal by restoring the required complementary coverage.

Remark 3. *Ladder standard contradictions are maximal under literal extension: once additional literals are introduced, the resulting clause set generally fails to remain a standard contradiction. In particular, the following three types of extensions will be considered:*

(1) Add $\neg p_1$ to C_i for $i = 3, \ldots, m+1$.

(2) Add p_m to C_i for $i = 1, \ldots, m-1$.

(3) Add a positive literal p_n $(n = 3, \ldots, m-1)$ to C_i for $i = 1, \ldots, m-2$.

We next show how a clause set admitting any of the above extensions can be augmented to recover the standard-contradiction property.

Proposition 3. *Let $S = \{C_1, C_2, \cdots, C_{m+1}\}$ be a ladder standard contradiction. Add one literal $\neg p_1$ to C_i ($i = 3, \cdots, m+1$), and get a new clause set S', then S' is a standard contradiction if and only if it satisfies*

(1) *If $i = 3$, then add $C_3' = p_1 \vee \neg p_2$.*

(2) *If $i > 3$, then add either $C_3' = p_1 \vee \neg p_2$ or $C_i' = p_1 \vee \neg p_{i-1}$.*

Proof. A ladder standard contradiction cannot remain a standard contradiction if additional literals are added.

After adding $\neg p_1$ to C_i ($i = 3, \cdots, m+1$), select a literal from each C_i ($i = 1, 2, \cdots, m+1$) to form a set. If we choose p_2 from C_1, p_j from C_j ($2 \leq j < i$), $\neg p_1$ from C_i, and $\neg p_{k-1}$ from C_k ($i < k \leq m+1$), then the selected literals form the set $\{p_2, p_2, \cdots, p_j, \cdots, p_{i-1}, \neg p_1, \cdots, \neg p_k, \cdots, \neg p_{m-1}, \neg p_m\}$. This selection contains no complementary pairs.

To ensure the presence of complementary pairs: If adding $C_3' = p_1 \vee \neg p_2$, selecting p_1 or $\neg p_2$ forms a complementary pair. If adding $C_i' = p_1 \vee \neg p_{i-1}$, then selecting p_1 or $\neg p_{i-1}$ forms a complementary pair. □

Proposition 4. *Let $S = \{C_1, C_2, \cdots, C_{m+1}\}$ be a ladder standard contradiction. Add $\neg p_1$ to C_3 and add b ($1 \leq b \leq i-2$) literals $\neg p_1$ to C_i ($i = 4, \cdots, m+1$), and get a new clause set S', then S' is a standard contradiction if it satisfies*

(1) *Add $C_3' = p_1 \vee p_2$.*

(2) *Add the clauses $n+1$ $C_3' = p_1 \vee p_2$ and $\bigwedge_i^n C_i'$, where $C_i' = p_1 \vee p_{i-1}$, ($0 \leq n \leq b$).*

Proof. The proof is similar to Proposition 3. □

Proposition 5. *Let $S = \{C_1, C_2, \cdots, C_{m+1}\}$ be a ladder standard contradiction. Add b ($1 \leq b \leq i-2$) literals $\neg p_1$ only to C_i ($i = 4, \cdots, m+1$), and get a new clause set S', then S' is a standard contradiction if it satisfies*

(1) *Add one clause $C_3' = p_1 \vee p_2$;*

(2) *Add a clauses $\bigwedge_{i=1}^a C_i'$, where $C_i' = p_1 \vee p_{i-1}$;*

(3) *Add $n+1$ clauses $C_3' = p_1 \vee p_2$ and $\bigwedge_{i=1}^n C_i'$, where $C_i' = p_1 \vee p_{i-1}$ ($0 \leq n \leq a$).*

Proof. The proof is similar to Proposition 3. □

Proposition 6. *Let $S = \{C_1, C_2, \cdots, C_{m+1}\}$ be a ladder standard contradiction. Add one p_m to C_i $(i = 1, \cdots, m-1)$, and get a new clause set S', then S' is a standard contradiction if and only if it satisfies*

(1) *If $i = m-1$, then add the clause $C'_{m-1} = \neg p_m \vee p_{m-1}$.*

(2) *If $i < m-1$, then add the clause $C'_{m-1} = \neg p_m \vee p_{m-1}$ or $C'_i = \neg p_m \vee p_i$.*

Proof. A ladder standard contradiction cannot remain a standard contradiction if additional literals are added.

After adding p_m to C_i $(i = 1, \cdots, m-1)$, choose one literal from each clause of C_i $(i = 1, 2, \cdots, m+1)$ to form a set. If $\neg p_{m-1}$ is chosen from clause C_{m+1}, $\neg p_{j-1}$ is chosen from C_j $(i < j \leq m)$, the added p_m is chosen from C_i, and p_k is chosen from C_k $(1 \leq k < i)$, then the set formed by the selected literals is $\{p_1, p_2, \cdots, p_k, \cdots, p_m, \neg p_i, \cdots, \neg p_{j-1}, \cdots, \neg p_{m-1}, \neg p_{m-1}\}$. Only when selecting in this way does the set contain no complementary pairs.

Adding $C'_{m-1} = \neg p_m \vee p_{m-1}$, choosing either $\neg p_m$ or p_{m-1} forms a complementary pair; Adding $C'_i = \neg p_m \vee p_i$, choosing either $\neg p_m$ or p_i forms a complementary pair. □

Proposition 7. *Let $S = \{C_1, C_2, \cdots, C_{m+1}\}$ be a ladder standard contradiction. Add p_m to C_{m-1} and add b $(1 \leq b \leq i-2)$ literals p_m to C_i $(i = 1, \cdots, m-2)$, and get a new clause set S', then S' is a standard contradiction if it satisfies*

(1) *Add $C'_{m-1} = \neg p_m \vee p_{m-1}$.*

(2) *Add $n+1$ clauses $C'_{m-1} = \neg p_m \vee p_{m-1}$ and $\bigwedge_i^n C'_i$, where $C'_i = \neg p_m \vee p_i$, $(0 \leq n \leq b)$.*

Proof. The proof is similar to Proposition 6. □

Proposition 8. *Let $S = \{C_1, C_2, \cdots, C_{m+1}\}$ be a ladder standard contradiction. Add b $(1 \leq b \leq i-2)$ literals p_m only to C_i $(i = 1, \cdots, m-2)$, and get a new clause set S', then S' is a standard contradiction if it satisfies*

(1) *Add one clause $C'_{m-1} = \neg p_m \vee p_{m-1}$;*

(2) *Add b clauses $\bigwedge_{i=1}^b C'_i$, where $C'_i = \neg p_m \vee p_i$;*

(3) *Add $n+1$ clauses $C'_{m-1} = \neg p_m \vee p_{m-1}$ and $\bigwedge_{i=1}^n C'_i$, where $C'_i = \neg p_m \vee p_i$, $(0 \leq n \leq b)$.*

Proof. The proof is similar to Proposition 6. □

Proposition 9. *Let $S = \{C_1, C_2, \cdots, C_{m+1}\}$ be a ladder standard contradiction. Add one p_n $(n = 3, \cdots, m-1)$ to C_i $(i = 1, \cdots, m-2)$ and get a new clause set S', then S' is a standard contradiction if and only if it satisfies*

(1) *If $i = 1, 2$, add the clauses $C_1' = C_2' = p_2 \vee \neg p_n$;*

(2) *If $i > 2$, add the clause $C_i' = p_i \vee \neg p_n$.*

Proof. A ladder standard contradiction cannot remain a standard contradiction if additional literals are added. After adding a positive literal p_n $(n = 3, \cdots, m-1)$ to C_i $(i = 1, \cdots, m-2)$, choose one literal from each clause of C_i $(i = 1, 2, \cdots, m+1)$ to form a set.

(1) Adding p_n $(n = 3, \cdots, m-1)$ to C_i $(i = 1, 2)$: If we select p_1 and p_n or p_n and $\neg p_1$ from clauses C_i $(i = 1, 2)$, select p_j from clauses C_j $(3 \leq j < n)$, select p_n from clause C_n, and select negated literals other than $\neg p_j$, $\neg p_n$ from clauses C_k $(n < k \leq m+1)$, then the set formed by the selected literals is $\{p_1, p_n, \cdots, p_j, p_n, \cdots, \neg p_k, \ldots\}$ or $\{p_n, \neg p_1, \cdots, p_j, p_n, \cdots, \neg p_k, \ldots\}$. Only when selecting in this way does the set contain no complementary pairs. Adding the clauses $C_1' = C_2' = p_2 \vee \neg p_n$, choosing either p_2 or $\neg p_n$ forms a complementary pair.

(2) Adding p_n $(n = 3, \cdots, m-1)$ to C_i $(i > 2)$: If we select p_j from clauses C_j $(1 < j < i)$, select p_n from clause C_i, select $\neg p_i$ from clauses C_k $(i < k < n)$, select p_n from clause C_n, and select negated literals other than $\neg p_j$, $\neg p_n$ from clauses C_l $(n < l \leq m+1)$, then the set formed by the selected literals is $\{p_1, \cdots, p_j, \cdots, p_{i-1}, p_n, \neg p_i, \cdots, \neg p_i, p_n, \cdots, \neg p_k, \ldots\}$. Only when selecting in this way does the set contain no complementary pairs. Adding clause $C_i' = p_i \vee \neg p_n$, choosing p_i or $\neg p_n$ forms a complementary pair.

□

Proposition 10. *Let $S = \{C_1, C_2, \cdots, C_{m+1}\}$ be a ladder standard contradiction. Add any number of p_n $(n = 3, \cdots, m-1)$ to C_i $(i = 1, \cdots, m-2)$ and get a new clause set S', then S' is a standard contradiction if it satisfies*

(1) *If $i = 1, 2$, add the clauses $C_{1n}' = C_{2n}' = p_2 \vee \neg p_n$;*

(2) *If $i > 2$, add the clause $C_{in}' = p_i \vee \neg p_n$.*

Proof. The proof is similar to Proposition 9. □

If C_1 and C_2 are extended by the same positive literal, only one additional clause is required. Suppose C_1 is extended by a literals, C_2 by b literals, and each C_i $(i = 3, \ldots, m-2)$ by c literals. Then the number of added clauses satisfies $c < N_{\text{add}} \leq a + b + c$.

Remark 4. *The clauses introduced via Proposition 10 do not necessarily constitute a standard contradiction on their own. For instance, let* $S = \{C_1, C_2, \ldots, C_6\}$ *be a ladder standard contradiction. If* p_3 *and* p_4 *are added to both* C_1 *and* C_2*, then maintaining the standard-contradiction property requires adding* $C'_{13} = C'_{23} = p_2 \vee \neg p_3, C'_{14} = C'_{24} = p_2 \vee \neg p_4$*. However, the added-clause set (e.g.,* $\{C'_{13}, C'_{14}\}$*) contains only negated added literals. Consequently, selecting one added literal from each added clause yields no complementary pair, and hence the added clauses alone fail to form a standard contradiction.*

Proposition 11. *Let* $S = \{C_1, C_2, \cdots, C_{m+1}\}$ *be a ladder standard contradiction. Add any number of* $\neg p_1$ *to* C_i $(i = 3, \cdots, m+1)$ *and any number of* p_n $(n = 3, \cdots, m-1)$ *to* C_j $(j = 1, \cdots, m-2)$*, and get a new set of clause set* S'*. If adding clauses according to Propositions 4 and 10, then* S' *is a standard contradiction.*

Proof. The conclusion follows from Propositions 4 and 10. □

Proposition 12. *Let* $S = \{C_1, C_2, \cdots, C_{m+1}\}$ *be a ladder standard contradiction. Add any number of* p_m *to* C_i $(i = 1, \cdots, m-1)$ *and any number of* p_n $(n = 3, \cdots, m-1)$ *to* C_j $(j = 1, \cdots, m-2)$*, and get a new set of clause set* S'*. If adding clauses according to Propositions 7 and 10, then* S' *is a standard contradiction.*

Proof. The conclusion follows from Propositions 7 and 10. □

Proposition 13. *Let* $S = \{C_1, C_2, \cdots, C_{m+1}\}$ *be a ladder standard contradiction. Add one* $\neg p_1$ *to* C_i $(i = 3, \cdots, m+1)$ *and one* p_m *to* C_j $(j = 1, \cdots, m-1)$*, and get a new clause set* S'*, then* S' *is a standard contradiction if it satisfies*

(1) *Add the clauses either* $C'_3 = p_1 \vee \neg p_2$ *or* $C'_i = p_1 \vee \neg p_{i-1}$,

(2) *Add the clauses either* $C'_{m-1} = \neg p_m \vee p_{m-1}$ *or* $C'_j = \neg p_m \vee p_j$,

where $i = m+1$, $j = 1$, $C'_i = C'_j = p_1 \vee \neg p_m$.

Proof. Add $\neg p_1$ and p_m sequentially, the conclusion follows by Propositions 3 and 6. □

Proposition 14. *Let $S = \{C_1, C_2, \cdots, C_{m+1}\}$ be a ladder standard contradiction. Add a literals $\neg p_1$ to C_i $(i = 3, \cdots, m+1)$ and b literals p_m to C_j $(j = 1, \cdots, m-1)$, and get a new clause set S', then S' is a standard contradiction if it satisfies*

(1) *Add $C_3' = p_1 \vee \neg p_2$ and $\bigwedge_{j=1}^{b} C_j'$, where $C_j' = \neg p_m \vee p_j$, $(1+b)$ clauses in total,*

(2) *Add $\bigwedge_{i=1}^{a} C_i'$ and $C_{m-1}' = \neg p_m \vee p_{m-1}$, where $C_i' = p_1 \vee \neg p_{i-1}$, $(1+a)$ clauses in total,*

(3) *Add $\bigwedge_{i=1}^{a} C_i'$ and $\bigwedge_{j=1}^{b} C_j'$, where $C_i' = p_1 \vee \neg p_{i-1}$ and $C_j' = \neg p_m \vee p_j$, $(a+b-1)$ clauses in total,*

where i and j are the sequence numbers of the clause with the added literal.

Proof. Add multiple $\neg p_1$ and multiple p_m sequentially. The conclusion follows from Propositions 3 and 6. □

Remark 5. *The clauses introduced in Proposition 14 do not, in general, form a standard contradiction by themselves. For example, let $S = \{C_1, C_2, \ldots, C_6\}$ be a ladder standard contradiction. If we further add $\neg p_1$ to C_5, add p_5 to C_2, and add p_3, p_4 to C_1, then preserving the standard-contradiction property requires adding*

$$C_5' = p_1 \vee \neg p_4, C_2' = \neg p_5 \vee p_2, C_{13}' = p_2 \vee \neg p_3, C_{14}' = p_2 \vee \neg p_4.$$

Nevertheless, the added-clause set $\{C_5', C_2', C_{13}', C_{14}'\}$ is not a standard contradiction.

Remarks 3 and 4 are meant to clarify the underlying structural mechanism. In particular, the contradiction is obtained from the interaction between the original clause set and the added auxiliary ("patch") clauses; the auxiliary clauses are not required to form a standard contradiction on their own. Accordingly, literal extensions should be understood as local modifications that may violate the original complementary coverage, and the added clauses serve to repair these local violations while preserving the global contradiction pattern.

4 Composition and construction of dual-line standard contradictions

This section studies how two standard contradictions can be combined via clause connections. We characterize when the connected clause set is again a standard contradiction and give a constructive scheme for forming such compositions. For a clause set $\bigwedge_{i=1}^{m} C_i$, we write $(\bigwedge_{i=1}^{m} C_i) \setminus C_i$ for the conjunction obtained by deleting C_i $(1 \leq i \leq m)$.

4.1 Multiple-clause composition

We consider compositions that use several connecting clauses between two dual-line standard contradictions and determine when the resulting set remains a standard contradiction. When a direct composition fails, we provide repair conditions and distinguish cases with and without repeated literals.

Remark 6. *Let two dual-line standard contradictions* $S_1 = \bigwedge_{i=1}^{m+1} C_i$ *and* $S_2 = \bigwedge_{j=1}^{n+1} B_j$ *have no repeated literals, where* $D_1 = C_i \cup B_k$, $D_2 = C_j \cup B_l$, *and* $i, j \in [1, m+1]$, $i \neq j$, $k, l \in [1, n+1]$, $k \neq l$. *Then* $(S_1 \backslash (C_i \wedge C_j)) \wedge (S_2 \backslash (B_k \wedge B_l)) \wedge D_1 \wedge D_2$ *does not form a standard contradiction.*

Remark 7. *If two dual-line standard contradictions* $S_1 = \bigwedge_{i=1}^{m+1} C_i$ *and* $S_2 = \bigwedge_{j=1}^{n+1} B_j$ *have no repeated literals, when each has a* $(a \geq 2)$ *clauses participating in the composition, let* $D_k = C_{i_k} \cup B_{j_k}$, $i \in [1, m+1]$, $j \in [1, n+1]$, $k \in [1, a]$. *Then the set of composed clauses* $(S_1 \setminus (C_i \wedge C_j)) \wedge (S_2 \setminus (B_k \wedge B_l)) \wedge D_1 \wedge \cdots \wedge D_a$ *does not form a standard contradiction.*

Proposition 15. *Suppose that two dual-line standard contradictions* $S_1 = \bigwedge_{i=1}^{m+1} C_i$ *and* $S_2 = \bigwedge_{j=1}^{n+1} B_j$ *do not have repeated literals, where* $D_1 = C_i \cup B_k$, $D_2 = C_j \cup B_l$, *and* $i, j \in [1, m+1]$, $i \neq j$, $k, l \in [1, n+1]$, $k \neq l$. *Then* $(S_1 \setminus (C_i \wedge C_j)) \wedge (S_2 \setminus (B_k \wedge B_l)) \wedge D_1 \wedge D_2 \wedge E_\alpha \wedge E_\beta$ *is a standard contradiction if* E_α, E_β *are any two of* C_i, B_k, C_j, B_l.

Proof. According to Remark 6, $(S_1 \setminus (C_i \wedge C_j)) \wedge (S_2 \setminus (B_k \wedge B_l)) \wedge D_1 \wedge D_2$ is not a standard contradiction because neither the set $\{p_1, p_2, \cdots, p_{i-1}, p_{i+1}, \cdots, p_{j-1}, p_{j+1}, \cdots, p_{m+1}, r_1\}$ nor $\{q_1, q_2, \cdots, q_{k-1}, q_{k+1}, \cdots, q_{l-1}, q_{l+1}, \cdots, q_{n+1}, r_2\}$ contains complementary pairs. Adding clauses E_α and E_β, there are five cases.

(1) If E_α, E_β are C_i, C_j or B_k, B_l, then the new set of clauses contains S_1 or S_2, which can constitute a standard contradiction with redundant clauses.

(2) If E_α, E_β are C_i, B_k, the new set of clauses is equivalent to $(S_1 \setminus C_j) \wedge (S_2 \setminus B_l) \wedge D_1 \wedge D_2$, according to the conclusion of the single-clause composition, it can form a standard contradiction with the redundant clause D_1.

(3) If E_α, E_β are C_j, B_l, the set of new clauses is equivalent to $(S_1 \setminus C_i) \wedge (S_2 \setminus B_k) \wedge D_1 \wedge D_2$, which can form a standard contradiction with the redundant clause D_2.

(4) If E_α, E_β are C_i, B_l, the set of new clauses is equivalent to $(S_1 \setminus C_j) \wedge (S_2 \setminus B_k) \wedge D_1 \wedge D_2$, which can form a standard contradiction without redundant clauses.

(5) If E_α, E_β are C_j, B_k, the set of new clauses is equivalent to $(S_1 \setminus C_i) \wedge (S_2 \setminus B_l) \wedge D_1 \wedge D_2$, which can form a standard contradiction without redundant clauses.

□

Proposition 16. *Suppose two dual-line standard contradictions $S_1 = \bigwedge_{i=1}^{m+1} C_i$ and $S_2 = \bigwedge_{j=1}^{n+1} B_j$, when each has two clauses participating in the composition, where $D_1 = C_i \cup B_k$, $D_2 = C_j \cup B_l$, and $i, j \in [1, m+1]$, $i \neq j$, $k, l \in [1, n+1]$, $k \neq l$. Then $(S_1 \setminus (C_i \wedge C_j)) \wedge (S_2 \setminus (B_k \wedge B_l)) \wedge D_1 \wedge D_2$ is a standard contradiction if $C_i = B_k$ or $C_j = B_l$.*

Proof. We prove the conclusion for the case where $C_i = B_k$. Let $p_\alpha \in C_\alpha$, $\alpha \in [1, i) \cup (i, j) \cup (j, m+1]$, $q_\beta \in B_\beta$, $\beta \in [1, k) \cup (k, l) \cup (l, n+1]$, $r_1 \in D_1$, $r_2 \in D_2$. Since $C_i = B_k$, $r_1 \in C_i$ and $r_1 \in B_k$, if $r_2 \in C_j$, then the set $\{p_1, \cdots, p_{i-1}, p_{i+1}, \cdots, p_{j-1}, p_{j+1}, \cdots, p_{m+1}, r_1, r_2\}$ contains complementary pairs; if $r_2 \in B_l$, then the set $\{q_1, \cdots, q_{k-1}, q_{k+1}, \cdots, q_{l-1}, q_{l+1}, \cdots, q_{n+1}, r_1, r_2\}$ contains complementary pairs. Hence, $\{p_1, \cdots, p_{i-1}, p_{i+1}, \cdots, p_{j-1}, p_{j+1}, \cdots, p_{m+1}, q_1, \cdots, q_{k-1}, q_{k+1}, \cdots, q_{l-1}, q_{l+1}, \cdots, q_{n+1}, r_1, r_2\}$ contains complementary pairs, making the composed clause set a standard contradiction. □

Proposition 17. *Suppose that there are two dual line standard contradictions $S_1 = \bigwedge_{i=1}^{m+1} C_i$ and $S_2 = \bigwedge_{j=1}^{n+1} B_j$, each having a ($a \geq 2$) clauses participating in the composition, where $D_k = C_{i_k} \cup B_{j_k}$, and $i \in [1, m+1]$, $j \in [1, n+1]$, $k \in [1, a]$. Then $(S_1 \setminus (C_{i_1} \wedge \cdots \wedge C_{i_a})) \wedge (S_2 \setminus (B_{j_1} \wedge \cdots \wedge B_{j_a})) \wedge D_1 \wedge \cdots \wedge D_a$ is a standard contradiction, if at least $a-1$ of $C_{i_k} = B_{j_k}$,*

Proof. Let $p_\alpha \in C_\alpha$, $\alpha \in [1, m+1]$, $q_\beta \in B_\beta$, $\beta \in [1, n+1]$; let $r_\gamma \in D_\gamma$, $\gamma \in [1, a]$. Since $C_{i_1} = B_{j_1}, \cdots, C_{i_{a-1}} = B_{j_{a-1}}$, $r_1 \in C_{i_1}$ and $r_1 \in B_{j_1}, \cdots, r_{a-1} \in C_{i_{a-1}}$ and $r_{a-1} \in B_{j_{a-1}}$. If $r_a \in C_{i_a}$, then the set $\{p_1, \cdots, p_{i_1}, \cdots, p_{i_{a-1}}, \cdots, p_{m+1}, r_1, \cdots, r_a\}$ contains complementary pairs; if $r_a \in B_{j_a}$, then the set $\{q_1, \cdots, q_{j_1}, \cdots, q_{j_{a-1}}, \cdots, q_{n+1}, r_1, \cdots, r_a\}$ contains complementary pairs. Therefore, $\{p_1, \cdots, p_{i_1}, \cdots, p_{i_{a-1}}, \cdots, p_{m+1}, q_1, \cdots, q_{j_1}, \cdots, q_{j_{a-1}}, \cdots, q_{n+1}, r_1, \cdots, r_a\}$ contains complementary pairs, making the set of composed clauses a standard contradiction. □

Proposition 18. *Suppose two dual-line standard contradictions $S_1 = \bigwedge_{i=1}^{m+1} C_i$ and $S_2 = \bigwedge_{j=1}^{n+1} B_j$, when each has two clauses participating in the composition, where $D_1 = C_i \cup B_k$, $D_2 = C_j \cup B_l$, and $i, j \in [1, m+1]$, $i \neq j$, $k, l \in [1, n+1]$, $k \neq l$. Then $(S_1 \setminus (C_i \wedge C_j)) \wedge (S_2 \setminus (B_k \wedge B_l)) \wedge D_1 \wedge D_2$ is a standard contradiction if it satisfies one of the following conditions.*

(1) *There exist* C_α, C_β *such that* $B_k = C_\alpha$, $B_l = C_\beta$.

(2) *There exist* B_α, B_β *such that* $C_i = B_\alpha$, $C_j = B_\beta$.

Proof. We prove the conclusion for the case where there exist C_α, C_β, such that $B_k = C_\alpha$, $B_l = C_\beta$. When $B_k = C_\alpha$, $B_l = C_\beta$, the set of clauses $\{B_1, \cdots, B_{k-1}, B_k, B_{k+1}, \cdots, B_{l-1}, B_l, B_{l+1}, \cdots, B_{n+1}\} = \{B_1, \cdots, B_{k-1}, C_\alpha, B_{k+1}, \cdots, B_{l-1}, C_\beta, B_{l+1}, \cdots, B_{n+1}\}$, the set of composed clauses contains the standard contradiction S_2, which makes it a standard contradiction. The case where there exist B_α, B_β such that $C_i = B_\alpha$, $C_j = B_\beta$ can be proven similarly. □

Proposition 19. *Suppose that two dual-line standard contradictions* $S_1 = \bigwedge_{i=1}^{m+1} C_i$ *and* $S_2 = \bigwedge_{j=1}^{n+1} B_j$, *a is a positive integer, when each has a* $(2 \leq a < (\min\{m+1, n+1\}/2))$ *clauses participating in the composition, where* $D_k = C_{i_k} \cup B_{j_k}$, $i \in [1, m+1]$, $j \in [1, n+1]$, $k \in [1, a]$. *Then the composed clause set* $(S_1 \setminus (C_{i_1} \wedge \cdots \wedge C_{i_a})) \wedge (S_2 \setminus (B_{j_1} \wedge \cdots \wedge B_{j_a})) \wedge D_1 \wedge \cdots \wedge D_a$ *is a standard contradiction if at least a of* $C_\gamma = B_{j_k}$ *or* $B_\gamma = C_{i_k}$, $\gamma \in [1, a]$,

Proof. The conclusion follows by mathematical induction on Proposition 18. □

In the symmetric case $m = n$, if each dual-line standard contradiction contributes $\frac{m+1}{2}$ clauses to the connection, then the resulting clause set is a standard contradiction only in the degenerate situation where the two inputs coincide, i.e., $S_1 = S_2$.

Accordingly, we impose explicit upper bounds on the number a of clauses participating in a composition. Beyond these bounds, one can select one literal per clause so that no complementary pair occurs, and the standard-contradiction property is lost.

4.2 Generation algorithms

We next present constructive generation schemes for dual-line and triple-line standard contradictions and count the number of distinct structures obtainable for a fixed number of literals.

Theorem 4. *When the number of literals (with p and* $\neg p$ *considered as one type) is fixed at k, the number of dual-line standard contradictions that can be generated is* $(k!/2)$ *in Algorithm 1.*

Theorem 5. *When the number of literals (with p and* $\neg p$ *considered as one type) is fixed at k, the number of standard triple-line contradictions generated is* $\binom{k}{3}$ *in Algorithm 2.*

Algorithm 1: A method for generating dual-line standard contradictions

```
Input: Set of literals {p_1, p_2, ..., p_k} (k ≥ 3).
Output: Dual-line standard contradiction S.
Initiate S;
Perform a full permutation of the elements in the literal set and sequentially store them in S[j];
for j = 1 to (k!) do
    for i = 1 to k + 1 do
        if i = 1 then
            S[j][i] ← p_i ∨ p_{i+1};
        end
        if 2 ≤ i ≤ k then
            S[j][i] ← ¬p_{i-1} ∨ p_i;
        end
        if i = k + 1 then
            S[j][i] ← ¬p_{i-2} ∨ ¬p_{i-1};
        end
    end
end
return S;
```

Algorithm 2: A method for generating triple-line standard contradictions

```
Input: Set of literals {p_1, p_2, ..., p_k} (k ≥ 3)
Output: Triple-line standard contradiction S
Initialize S as an empty set;
Perform a full permutation of the elements in the literal set and sequentially store them in S[j];
for j = 1 to (k!) do
    for i = 1 to k do
        if 1 ≤ i < k - 1 then
            S[i] ← p_i ∨ p_{i+1} ∨ p_{i+2};
            S[i + 1] ← p_i ∨ ¬p_{i+1} ∨ p_{i+2};
            S[i + 2] ← ¬p_i ∨ p_{i+1} ∨ p_{i+2};
            S[i + 3] ← ¬p_i ∨ ¬p_{i+1} ∨ p_{i+2};
        end
        if i = k then
            S[4i - 3] ← p_{i-1} ∨ p_i ∨ ¬p_{i+1};
            S[4i - 2] ← p_{i-1} ∨ ¬p_i ∨ ¬p_{i+1};
            S[4i - 1] ← ¬p_{i-1} ∨ p_i ∨ ¬p_{i+1};
            S[4i] ← ¬p_{i-1} ∨ ¬p_i ∨ ¬p_{i+1};
        end
    end
end
return S;
```

The above algorithms are not designed to guide or accelerate the proof search of concrete automated theorem provers. They serve as constructive witnesses for the existence and generality of the clause-set patterns studied here. In any practical integration, one would not enumerate permutations; instead, one would rely on lightweight syntactic matching and the prover's existing indexing infrastructure to detect small instances of the templates as candidates for CS/S-CS macro-steps. Developing and evaluating such detection heuristics is outside the scope of this paper.

5 Standard contradictions in first-order logic

In first-order logic, a clause set does not, in general, constitute a standard contradiction in isolation: substitutions are essential. Nevertheless, by repeatedly reusing certain clauses and applying suitable substitutions, one can derive a new clause set that forms a standard contradiction. This section presents several representative clause families that are not standard contradictions initially but can be transformed into standard contradictions via clause reuse and appropriate substitutions.

Proposition 20. *Let* $S = \{C_1, C_2, \cdots, C_{n+1}\} = P(t_{11}, \cdots, a_1, \cdots, t_{1m}) \wedge \cdots \wedge \neg P(t_{g1}, \cdots, a_g, \cdots, t_{gm}) \wedge \cdots \wedge (\neg P(t'_{11}, \cdots, t'_{1m}) \vee P(t'_{21}, \cdots, t'_{2m}) \vee \cdots \vee P(t'_{(n+1)1}, \cdots, t'_{(n+1)m}))$ $(2 \leq g \leq n)$ *in first-order logic, where* P *is a m-ary predicate symbol, and* t_i *are terms. Then* S *can form a standard contradiction by repeating clause* C_n *once if it satisfies*

(1) $t_{hi} \in C_h$ *is a constant symbol* a_h $(1 \leq h \leq n, 1 \leq i \leq m)$.

(2) t'_{ki} *and* $t'_{(k-1)j} \in C_{n+1}$ *are the same variable symbols* $(1 < k \leq n)$, $t'_{(n-1)i}$ *and* t'_{nj} *are the same variable,* $t'_{(n+1)i}$ *and* $t'_{(n+1)j}$ *are the same variable symbols as* t'_{ni} *(or* $t'_{(n-1)j}$*)* $(1 \leq j \leq m, j \neq i)$.

(3) t'_{1i} *is any variable symbol different from the variable symbols above.*

C_1	C_2	$\cdots$	C_n	C_{n+1}
		$\cdots$		$P(t'_{(n+1)1}, \ldots, \ldots, t'_{(n+1)m})$
		$\cdots$		$\cdots$
		$\cdots$		$P(t'_{21}, \ldots, \ldots, t'_{2m})$
$P(t_{11}, \ldots, a_1, \ldots, t_{1m})$	$\neg P(t_{21}, \ldots, a_2, \ldots, t_{2m})$	$\cdots$	$\neg P(t_{n1}, \ldots, a_n, \ldots, t_{nm})$	$\neg P(t'_{11}, \ldots, \ldots, t'_{1m})$

Table 3: A clause set with the structure of Proposition 20

Proof. To ensure that C_{n+1} contains complementary pairs for C_1, substitute a_1/t'_{1i}; and so on until C_n, substitute a_n/t'_{ni}. From the condition that t'_{ki} and $t'_{(k-1)j}$ are the same variable symbols, then $a_2/t'_{1j}, \cdots, a_n/t'_{(n-1)j'}$; From the condition that $t'_{(n-1)i}$ and t'_{nj} are the same variable symbols, then $a_{(n-1)}/t'_{nj}$; From the condition that

$t'_{(n+1)i}$ and $t'_{(n+1)j}$ are the same variable symbols as t'_{ni} (or $t'_{(n-1)j}$), then $a_n/t'_{(n+1)i}$, $a_n/t'_{(n+1)j}$.

According to these substitutions, the substitution for t_{1j} can only be a_2/t_{1j}, and so on until $t_{(n-1)j}$, it can only be $a_n/t_{(n-1)j}$, and for t_{nj} it can only be $a_{(n-1)}/t_{nj}$. Substitute t'_{kl} and t_{kl} with the same term $(1 \leq k \leq n, 1 \leq l \leq m)$, and substitute $t'_{(n+1)l}$ and t_{nl} with the same term $(1 \leq l \leq m)$. After performing these substitutions, the selection of atoms $C_{(n+1)_1}, \cdots, C_{(n+1)_n}$ in C_{n+1} will form complementary pairs, while the selection of $C_{(n+1)_{(n+1)}}$ will not form complementary pairs.

Therefore, it is necessary to repeat C_n and apply a_n/t_{nj} to satisfy the formation of complementary pairs for any selected atom in the clause set. □

We next present a concrete example illustrating the use of the above structure in S-CS-based dynamic deduction in first-order logic.

Example 2. *Let a clause set* $S = \{C_1, \ldots, C_7\}$ *in first-order logic, where* $C_1 = p(a_1, x_1) \vee p(a_4)$, $C_2 = \neg p(a_2, x_2) \vee p(a_5)$, $C_3 = \neg p(a_3, x_3) \vee p(a_5)$, $C_4 = \neg p(y_0, y_1) \vee p(y_1, y_2) \vee p(y_2, y_1) \vee p(y_2, y_2)$, $C_5 = \neg p(a_4) \vee p(a_6)$, $C_6 = \neg p(a_5) \vee p(a_6)$, $C_7 = \neg p(a_6)$.

Separated from $C_1 \sim C_4$: $C_1^- = p(a_1, x_1)$, $C_1^+ = p(a_4)$; $C_2^- = \neg p(a_2, x_2)$, $C_2^+ = p(a_5)$; $C_3^- = \neg p(a_3, x_3)$, $C_3^+ = p(a_5)$; $C_4^- = \neg p(y_0, y_1) \vee p(y_1, y_2) \vee p(y_2, y_1) \vee p(y_2, y_2)$, $C_4^+ = \Box$.

1. *According to Proposition 20,* $\bigwedge_{i=1}^4 C_i^-$ *is a standard contradiction, apply substitutions to* C_3*, resulting in* $C_3^{\sigma_1}$ *and* $C_3^{\sigma_2}$, $\sigma_1 = \{a_1/y_0, a_2/y_1, a_3/y_2, a_2/x_1, a_3/x_2, a_3/x_3\}$, $\sigma_2 = \{a_1/y_0, a_2/y_1, a_3/y_2, a_2/x_1, a_3/x_2, a_2/x_3\}$.

   Using the S-CS rule for the clauses C_1, C_2, C_3, C_4*, we obtain a S-CSC:* $C_8 = \mathcal{C}_5(C_1^{\sigma_1}, C_2^{\sigma_1}, C_3^{\sigma_1}, C_3^{\sigma_2}, C_4^{\sigma_1}) = P(a_4) \vee P(a_5)$.

2. *The second standard contradiction is* $(P(a_4) \vee P(a_5)) \wedge (\neg P(a_4) \vee P(a_6)) \wedge (\neg P(a_5) \vee P(a_6)) \wedge \neg P(a_6)$. *Using the CS rule,* $C_9 = \mathcal{C}_4(C_8, C_5, C_6, C_7) = \Box$.

The following shows binary resolution in multiple steps:
$C_8 = \mathcal{R}(C_3, C_4) = \neg p(y_0, y_1) \vee p(y_1, y_2) \vee p(y_2, y_1)$
$C_9 = \mathcal{R}(C_3, C_8) = \neg p(y_0, y_1) \vee p(y_1, y_2)$
$C_{10} = \mathcal{R}(C_1, C_9) = p(y_1, y_2) \vee p(a_4)$
$C_{11} = \mathcal{R}(C_2, C_{10}) = p(a_4) \vee p(a_5)$
$C_{12} = \mathcal{R}(C_5, C_{11}) = p(a_5) \vee p(a_6)$
$C_{13} = \mathcal{R}(C_6, C_{12}) = p(a_6)$
$C_{14} = \mathcal{R}(C_7, C_{13}) = \Box$
This shows that binary resolution requires 7 steps to derive □, whereas the S-CS-based dynamic deduction above reaches □ in 2 steps.

Proposition 21. *Let* $S = \{C_1, C_2, \cdots, C_{n+1}\} = P(t_{r1}, \cdots, a_r, \cdots, t_{rm}) \wedge \cdots \wedge \neg P(t_{s1}, \cdots, a_s, \cdots, t_{sm}) \wedge \cdots \wedge (\neg P(t'_{11}, \cdots, t'_{1m}) \vee \cdots \vee \neg P(t'_{r1}, \cdots, t'_{rm}) \vee P(t'_{(r+1)1}, \cdots, t'_{(r+1)m}) \vee \cdots \vee P(t'_{(n+1)1}, \cdots, t'_{(n+1)m}))$ *in first-order logic* $(1 \le r < n, r < s \le n)$*, where* P *is any* m*-ary predicate symbol, and* t_i *are terms. Then* S *can form a standard contradiction by repeating clause* C_n *once if it satisfies*

(1) $t_{hi} \in C_h$ *is a constant symbol* a_h $(1 \le h \le n, 1 \le i \le m)$.

(2) t'_{ki} *and* $t'_{(k-1)j} \in C_{n+1}$ *are the same variable symbols* $(1 < k \le n)$, $t'_{(n-1)i}$ *and* t'_{nj} *are the same variable,* $t'_{(n+1)i}$ *and* $t'_{(n+1)j}$ *are the same variable symbols as* t'_{ni} *(or* $t'_{(n-1)j}$*)* $(1 \le j \le m, j \ne i)$.

(3) t'_{1i} *is any variable symbol different from the variable symbols above.*

Proof. To ensure that C_{n+1} contains complementary pairs for C_1, substitute a_1/t'_{1i}; and so on until C_n, substitute a_n/t'_{ni}. From the condition that t'_{ki} and $t'_{(k-1)j}$ are the same variable symbols, then $a_2/t'_{1j}, \cdots, a_n/t'_{(n-1)j'}$; From the condition that $t'_{(n-1)i}$ and t'_{nj} are the same variable symbols, then $a_{(n-1)}/t'_{nj}$; From the condition that $t'_{(n+1)i}$ and $t'_{(n+1)j}$ are the same variable symbols as t'_{ni} (or $t'_{(n-1)j}$), then $a_n/t'_{(n+1)i}$, $a_n/t'_{(n+1)j}$.

According to these substitutions, the substitution for t_{1j} can only be a_2/t_{1j}, and so on until $t_{(n-1)j}$, it can only be $a_n/t_{(n-1)j}$, and for t_{nj} it can only be $a_{(n-1)}/t_{nj}$. Substitute t'_{kl} and t_{kl} with the same term $(1 \le k \le n, 1 \le l \le m)$, and substitute $t'_{(n+1)l}$ and t_{nl} with the same term $(1 \le l \le m)$. After performing these substitutions, the selection of atoms $C_{(n+1)_1}, \cdots, C_{(n+1)_n}$ in C_{n+1} will form complementary pairs, while the selection of $C_{(n+1)_{(n+1)}}$ will not form complementary pairs.

Therefore, it is necessary to repeat C_n and apply a_n/t_{nj} to satisfy the formation of complementary pairs for any selected atom in the clause set. □

Several cases are analogous to Proposition 21 but lead to different outcomes, including situations in which no clause repetition is required to obtain a standard contradiction and situations in which a standard contradiction cannot be formed.

Proposition 22. *Let* $S = \{C_1, C_2, C_3\} = (P(t_{11}, \cdots, a_1, \cdots, t_{1m}) \vee P(t_{21}, \cdots, a_2, \cdots, t_{2m}) \vee \cdots \vee P(t_{(n-1)1}, \cdots, a_{(n-1)}, \cdots, t_{(n-1)m})) \wedge (\neg P(t_{n1}, \cdots, a_n, \cdots, t_{nm})) \wedge (\neg P(t'_{11}, \cdots, t'_{1m}) \vee P(t'_{21}, \cdots, t'_{2m}) \vee \cdots \vee P(t'_{(n+1)1}, \cdots, t'_{(n+1)m}))$ *in first-order logic, where* P *is any* m*-ary predicate symbol, and* t_i *are terms. Then* S *can form a standard contradiction if it satisfies*

(1) $t_{hi} \in C_1$ *is a constant symbol* a_h $(1 \le h \le n-1)$*, and in* C_2*,* t_{ni} *is a constant symbol* a_n *for* $(1 \le i \le m)$.

(2) t'_{ki} *and* $t'_{(k-1)j} \in C_3$ *are the same variable symbols* $(1 < k \leq n)$, $t'_{(n-1)i}$ *and* t'_{nj} *are the same variable,* $t'_{(n+1)i}$ *and* $t'_{(n+1)j}$ *are the same variable symbols as* t'_{ni} *(or* $t'_{(n-1)j}$*)* $(1 \leq j \leq m, j \neq i)$.

(3) t'_{1i} *is any variable symbol different from the variable symbols above.*

Proof. It suffices to ensure that C_3 contains complementary pairs for C_2. Consider making C_2 complementary to C_{3_n} and $C_{3_{(n+1)}}$ using a_n/t'_{ni}, a_n/t'_{nj}. From the condition that $t'_{(n+1)i}$ and $t'_{(n+1)j}$ are the same variable symbol as t'_{ni} (or $t'_{(n-1)j}$), then $a_n/t'_{(n+1)i}$, $a_n/t'_{(n+1)j}$.

After performing these substitutions, regardless of how the atoms are chosen in C_1, the selection of atoms C_{3_n} or $C_{3_{(n+1)}}$ in C_3 will form complementary pairs. □

Corollary 1. *Let* $S = \{C_1, C_2, C_3\} = (P(t_{11}, \cdots, a_1, \cdots, t_{1m}) \vee \cdots \vee P(t_{r1}, \cdots, a_r, \cdots, t_{rm}) \vee \neg P(t_{(r+1)1}, \cdots, a_{(r+1)}, \cdots, t_{(r+1)m}) \vee \neg P(t_{(n-1)1}, \cdots, a_{(n-1)}, \cdots, t_{(n-1)m}) \wedge (\neg P(t_{n1}, \cdots, a_n, \cdots, t_{nm})) \wedge (\neg P(t'_{11}, \cdots, t'_{1m}) \vee \cdots \vee \neg P(t'_{r1}, \cdots, t'_{rm}) \vee P(t'_{(r+1)1}, \cdots, t'_{(r+1)m}) \vee \cdots \vee P(t'_{(n+1)1}, \cdots, t'_{(n+1)m}))$ *in first-order logic* $(1 \leq r < n)$ *, where* P *is any* m*-ary predicate symbol, and* t_i *are terms. Then* S *can form a standard contradiction if it satisfies*

(1) $t_{hi} \in C_1$ *is a constant symbol* a_h $(1 \leq h \leq n-1)$*, and* $t_{ni} \in C_2$ *is a constant symbol* a_n $(1 \leq i \leq m)$.

(2) t'_{ki} *and* $t'_{(k-1)j} \in C_3$ *are the same variable symbols* $(1 < k \leq n)$, $t'_{(n-1)i}$ *and* t'_{nj} *are the same variable,* $t'_{(n+1)i}$ *and* $t'_{(n+1)j}$ *are the same variable symbols as* t'_{ni} *(or* $t'_{(n-1)j}$*)* $(1 \leq j \leq m, j \neq i)$.

(3) t'_{1i} *is any variable symbol different from the variable symbols above.*

Proof. It suffices to ensure that C_3 contains complementary pairs for C_2. Consider making C_2 complementary to C_{3_n} and $C_{3_{(n+1)}}$ using a_n/t'_{ni}, a_n/t'_{nj}. From the condition that $t'_{(n+1)i}$ and $t'_{(n+1)j}$ are the same variable symbols as t'_{ni} (or $t'_{(n-1)j}$), then $a_n/t'_{(n+1)i}$, $a_n/t'_{(n+1)j}$.

After performing these substitutions, regardless of how the atoms are chosen in C_1, the selection of atoms C_{3_n} or $C_{3_{(n+1)}}$ in C_3 will form complementary pairs. □

We extend the conclusions to n m-ary predicates as shown in Proposition 23.

Proposition 23. *Let* $S = \{C_1, C_2, \cdots, C_{n+1}\} = P_r(t_{r1}, \cdots, a_r, \cdots, t_{rm}) \wedge \cdots \wedge (\neg P_s(t_{s1}, \cdots, a_s, \cdots, t_{sm}) \wedge \ldots \wedge (\neg P_1(t'_{11}, \cdots, t'_{1m}) \vee \cdots \vee \neg P_r(t'_{r1}, \cdots, t'_{rm}) \vee$

$P_{r+1}(t'_{(r+1)1},\cdots,t'_{(r+1)m})\vee\cdots\vee P_n(t'_{n1},\cdots,t'_{nm})\vee P_n(t'_{(n+1)1},\cdots,t'_{(n+1)m}))$ in first-order logic ($1\leq r<n$, $r<s\leq n$), where $P_1,\cdots,P_n$ are n m-ary predicate symbols, and t_i are terms. Then S can form a standard contradiction by repeating clause C_n once if it satisfies

(1) *$t_{hi}\in C_h$ is a constant symbol a_h $(1\leq h\leq n, 1\leq i\leq m)$.*

(2) *t'_{ki} and $t'_{(k-1)j}\in C_{n+1}$ are the same variable symbols $(1<k\leq n)$, $t'_{(n-1)i}$ and t'_{nj} are the same variable symbols, $t'_{(n+1)i}$ and $t'_{(n+1)j}$ are the same variable symbols as t'_{ni} (or $t'_{(n-1)j}$) $(1\leq j\leq m, j\neq i)$.*

(3) *t'_{1i} is any variable symbol different from the variable symbols above.*

Proof. To ensure that C_{n+1} contains complementary pairs for C_1, substitute a_1/t'_{1i}; and so on until C_n, substitute a_n/t'_{ni}. From the condition that t'_{ki} and $t'_{(k-1)j}$ are the same variable symbols, then $a_2/t'_{1j},\cdots,a_n/t'_{(n-1)j'}$. From the condition that $t'_{(n-1)i}$ and t'_{nj} are the same variable symbols, we have $a_{(n-1)}/t'_{nj}$. From the condition that $t'_{(n+1)i}$ and $t'_{(n+1)j}$ are the same variable symbols as t'_{ni} (or $t'_{(n-1)j}$), then $a_n/t'_{(n+1)i}$, $a_n/t'_{(n+1)j}$.

According to these substitutions, the substitution for t_{1j} can only be a_2/t_{1j}, and so on until $t_{(n-1)j}$, it can only be $a_n/t_{(n-1)j}$, and for t_{nj} it can only be $a_{(n-1)}/t_{nj}$. Substitute t'_{kl} and t_{kl} with the same term $(1\leq k\leq n, 1\leq l\leq m)$, and substitute $t'_{(n+1)l}$ and t_{nl} with the same term $(1\leq l\leq m)$. After performing these substitutions, the selection of atoms $C_{(n+1)_1},\cdots,C_{(n+1)_n}$ in C_{n+1} will form complementary pairs, while the selection of $C_{(n+1)_{(n+1)}}$ will not form complementary pairs.

Therefore, it is necessary to repeat C_n and apply a_n/t_{nj} to satisfy the formation of complementary pairs for any selected atom in the clause set. □

Replacing constants and variables by function terms while preserving the same structural pattern yields Proposition 24.

Proposition 24. *Let $S=\{C_1,C_2,\cdots,C_{n+1}\}=P(t_{r1},\cdots,f_r(x),\cdots,t_{rm})\wedge\cdots\wedge(\neg P(t_{s1},\cdots, f_s(x),\cdots,t_{sm}))\wedge\cdots\wedge(\neg P(t'_{11},\cdots,t'_{1m})\vee\cdots\vee\neg P(t'_{r1},\cdots,t'_{rm})\vee P(t'_{(r+1)1},\cdots,t'_{(r+1)m})\vee\cdots\vee P(t'_{(n+1)1},\cdots,t'_{(n+1)m}))$ in the first order logic ($1\leq r<n, r<s\leq n$), where P is the m-ary predicate symbol, and t_i are terms. Then S can form a standard contradiction by repeating clause C_n once if it satisfies*

(1) *$t_{hi}\in C_h$ is a function symbol $f_h(x)$ $(1\leq h\leq n, 1\leq i\leq m)$.*

(2) *t'_{ki} and $t'_{(k-1)j}\in C_{n+1}$ are the same variable symbol $(1<k\leq n)$, $t'_{(n-1)i}$ and t'_{nj} are the same variable, $t'_{(n+1)i}$ and $t'_{(n+1)j}$ are the same variable symbol as t'_{ni} (or $t'_{(n-1)j}$) $(1\leq j\leq m, j\neq i)$.*

(3) t'_{1i} *is any variable symbol different from the variable symbols above.*

Where function terms $f_k(x)$ $(1 \leq k \leq n)$, x *can be any distinct constant, variable, or function term.*

Proof. Three cases exist for the conclusion.

(1) If x is a constant, the proof is identical to Proposition 21.

(2) If x is a variable, first substitute all variables x with a_0/x. To ensure that C_{n+1} contains complementary pairs for C_1, substitute $f_1(a_0)/t'_{1i}$; and so on until C_n, substitute $f_n(a_0)/t'_{ni}$. From the condition that t'_{ki} and $t'_{(k-1)j}$ are the same variable symbols, we have $f_2(a_0)/t'_{1j}, \cdots, f_n(a_0)/t'_{(n-1)j'}$. From the condition that $t'_{(n-1)i}$ and t'_{nj} are the same variable symbols, we have $f_{(n-1)}(a_0)/t'_{nj}$. From the condition that $t'_{(n+1)i}$ and $t'_{(n+1)j}$ are the same variable symbols as t'_{ni} (or $t'_{(n-1)j}$), then $f_n(a_0)/t'_{(n+1)i}$, $f_n(a_0)/t'_{(n+1)j}$.

According to these substitutions, the substitution for t_{1j} can only be $f_2(a_0)/t_{1j}$, and so on until $t_{(n-1)j}$, it can only be $f_n(a_0)/t_{(n-1)j}$, and for t_{nj} it can only be $f_{(n-1)}(a_0)/t_{nj}$. Substitute t'_{kl} and t_{kl} with the same term $(1 \leq k \leq n, 1 \leq l \leq m)$, and substitute $t'_{(n+1)l}$ and t_{nl} with the same term $(1 \leq l \leq m)$. After performing these substitutions, the selection of atoms $C_{(n+1)_1}, \cdots, C_{(n+1)_n}$ in C_{n+1} will form complementary pairs, while the selection of $C_{(n+1)_{(n+1)}}$ will not form complementary pairs.

Therefore, it is necessary to repeat C_n and apply $f_n(a_0)/t_{nj}$ to satisfy the formation of complementary pairs for any selected atom in the clause set.

(3) If x is a function term, perform first substitutions within the function terms, and the subsequent proof follows similarly to Proposition 21.

□

Corollary 2. *Let* $S = \{C_1, C_2, \cdots, C_{n+1}\}$ *in first-order logic,* $C_r = P(t_{r1}, \cdots, f_r(x), \cdots, t_{rm})$ *(or* $P(t_{r1}, \cdots, a_r, \cdots, t_{rm})$*)* $(1 \leq r < n)$, $C_s = \neg P(t_{s1}, \cdots, f_s(x), \cdots, t_{sm})$ *or* $P(t_{s1}, \cdots, a_s, \cdots, t_{sm})$ $(r < s \leq n)$, $C_{n+1} = \neg P(t'_{11}, \cdots, t'_{1m}) \vee \cdots \vee \neg P(t'_{r1}, \cdots, t'_{rm}) \vee P(t'_{(r+1)1}, \cdots, t'_{(r+1)m}) \vee \cdots \vee P(t'_{(n+1)1}, \cdots, t'_{(n+1)m})$, *where* P *is the* m*-ary predicate symbol, and* t_i *are terms. Then* S *can form a standard contradiction by repeating clause* C_n *once if it satisfies*

(1) $t_{hi} \in C_h$ *is either a constant symbol* a_h *or a function symbol* $f_h(x)$ $(1 \leq h \leq n, 1 \leq i \leq m)$.

(2) t'_{ki} *and* $t'_{(k-1)j} \in C_{n+1}$ *are the same variable symbols* $(1 < k \leq n)$, $t'_{(n-1)i}$ *and* t'_{nj} *are the same variables,* $t'_{(n+1)i}$ *and* $t'_{(n+1)j}$ *are the same variable symbols as* t'_{ni} *(or* $t'_{(n-1)j}$*)* $(1 \leq j \leq m, j \neq i)$.

(3) t'_{1i} *is any variable symbol different from the variable symbols above.*

Proof. The proof follows similarly to Proposition 24. □

Corollary 3. *Let* $S = \{C_1, C_2, \cdots, C_{n+1}\}$ *in first-order logic,* $C_r = P(t_{r1}, \cdots, f_r(n_1), \cdots, t_{rm})$ $(1 \leq r < n)$, $C_s = \neg P(t_{s1}, \cdots, f_s(n_1), \cdots, t_{sm})$ $(r < s \leq n)$, $C_{n+1} = \neg P(t'_{11}, \cdots, t'_{1m}) \vee \cdots \vee \neg P(t'_{r1}, \cdots, t'_{rm}) \vee P(t'_{(r+1)1}, \cdots, t'_{(r+1)m}) \vee \cdots \vee P(t'_{(n+1)1}, \cdots, t'_{(n+1)m})$, *where* P *is the symbol of the predicate* m*-ary, and* t_i *are terms. Then* S *can form a standard contradiction by repeating clause* C_n *once if it satisfies*

(1) $t_{hi} \in C_h$ *is a function symbol* $f_h(n_1)$ $(1 \leq h \leq n, 1 \leq i \leq m)$.

(2) t'_{ki} *and* $t'_{(k-1)j} \in C_{n+1}$ *are the same variable symbols* $(1 < k \leq n)$, $t'_{(n-1)i}$ *and* t'_{nj} *are the same variables,* $t'_{(n+1)i}$ *and* $t'_{(n+1)j}$ *are the same variable symbols as* t'_{ni} *(or* $t'_{(n-1)j}$*)* $(1 \leq j \leq m, j \neq i)$.

(3) t'_{1i} *is any variable symbol different from the variable symbols above.*

Where $f_k(n_1)$ *is an* n_1*-ary function symbol* $f_k(t_{ki1}, \cdots, t_{kin_1})$, $t_{ki1}, \cdots, t_{kin_1}$ *can be any variable symbol or function symbol under variables (*i *indicates the position in the* m*-ary predicate* P*)* $(1 \leq k \leq n)$.

Proof. The proof follows similarly to Proposition 24. □

Corollary 4. *Let* $S = \{C_1, C_2, \cdots, C_{n+1}\}$ *in first-order logic,* $C_r = P(t_{r1}, \cdots, f_r(n_1), \cdots, t_{rm})$ *or* $P(t_{r1}, \cdots, f(n_1), \cdots, t_{rm})$ $(1 \leq r < n)$, $C_s = \neg P(t_{s1}, \cdots, f_s(n_1), \cdots, t_{sm})$ *or* $P(t_{s1}, \cdots, f(n_1), \cdots, t_{sm})$ $(r < s \leq n)$, $C_{n+1} = \neg P(t'_{11}, \cdots, t'_{1m}) \vee \cdots \vee \neg P(t'_{r1}, \cdots, t'_{rm}) \vee P(t'_{(r+1)1}, \cdots, t'_{(r+1)m}) \vee \cdots \vee P(t'_{(n+1)1}, \cdots, t'_{(n+1)m})$, *where* P *is the* m*-ary predicate symbol, and* t_i *are terms. Then* S *can form a standard contradiction by repeating clause* C_n *once if it satisfies*

(1) $t_{hi} \in C_h$ *is a function symbol* $f_h(x)$ *or* $f(x)$ $(1 \leq h \leq n, 1 \leq i \leq m)$.

(2) t'_{ki} *and* $t'_{(k-1)j} \in C_{n+1}$ *are the same variable symbols* $(1 < k \leq n)$, $t'_{(n-1)i}$ *and* t'_{nj} *are the same variable symbols,* $t'_{(n+1)i}$ *and* $t'_{(n+1)j}$ *are the same variable symbols as* t'_{ni} *(or* $t'_{(n-1)j}$*)* $(1 \leq j \leq m, j \neq i)$.

(3) t'_{1i} *is any variable symbol different from the variable symbols above.*

Where $f_k(n_1)$ *is an* n_1*-ary function symbol* $f_k(t_{ki1}, \cdots, t_{kin_1})$*,* $f(n_1)$ *is an* n_1*-ary function symbol* $f(t_{ki1}, \cdots, t_{kin_1})$*,* $t_{ki1}, \cdots, t_{kin_1}$ *can be any variable symbol or function symbol under variables (*i *indicates the position in the symbol of the predicate* m *-ary* P*)* $(1 \le k \le n)$*.*

Proof. The proof follows similarly to Proposition 24. □

Proposition 25. *Let* $S = \{C_1, C_2, \cdots, C_{n+1}\} = (P(t_{r1}, \cdots, f(x_r), \cdots, t_{rm})) \wedge \cdots \wedge (\neg P(t_{s1}, \cdots, f(x_s), \cdots, t_{sm})) \wedge \cdots \wedge (\neg P(t'_{11}, \cdots, t'_{1m}) \vee \cdots \vee \neg P(t'_{r1}, \cdots, t'_{rm}) \vee P(t'_{(r+1)1}, \cdots, t'_{(r+1)m}) \vee \cdots \vee P(t'_{(n+1)1}, \cdots, t'_{(n+1)m}))$ *in first-order logic,* $(1 \le r < n, r < s \le n)$*, where* P *is the* m*-ary predicate symbol, and* t_i *are terms. Then* S *can form a standard contradiction if it satisfies*

(1) $t_{hi} \in C_h$ *is a function symbol* $f_h(x)$ $(1 \le h \le n, 1 \le i \le m)$*.*

(2) t'_{ki} *and* $t'_{(k-1)j} \in C_{n+1}$ *are the same variable symbols* $(1 < k \le n)$*,* $t'_{(n-1)i}$ *and* t'_{nj} *are the same variable symbols,* $t'_{(n+1)i}$ *and* $t'_{(n+1)j}$ *are the same variable symbols as* t'_{ni} *(or* $t'_{(n-1)j}$*)* $(1 \le j \le m, j \ne i)$*.*

(3) t'_{1i} *is any variable symbol different from the variable symbols above.*

Proof. In first-order logic, the same function in different variable symbols can be transformed into completely identical n terms by substituting all variable symbols with the same constant. Therefore, this structure can form a standard contradiction without repeating clauses. □

Remark 8. *In the case of Proposition 21, if* t'_{ki} *and* t'_{kj} *contain* $f(x)$ *or* $f_k(x)$*, then* S *cannot form a standard contradiction by repeating clause* C_n*.*

These results identify structural templates under which standard contradictions can be systematically generated via repeated clauses, provided that the involved terms satisfy mild conditions. The propositions cover both identical and distinct predicates, as well as constants, variables, and function terms, thereby providing a unified set of general criteria. We also isolate and classify several structurally similar configurations that nevertheless fail to yield standard contradictions.

6 Conclusion

We studied standard contradictions as clause-set patterns that can serve as cores for CS/S-CS macro-steps in clausal refutation. In the propositional case, we identified canonical non-unit families—dual-line, triple-line, and ladder—and analyzed how such cores behave under literal extension, auxiliary-clause patching, and clause-set composition. In the first-order case, we described representative families whose standard-contradiction property emerges only after suitable instantiation and controlled clause reuse. From the viewpoint of automated reasoning, the main point is structural rather than implementational: the results replace direct use of the combinatorial definition with explicit templates and conditions for recognizing candidate CS/S-CS cores. This provides a proof-theoretic basis for future work on core-detection heuristics and prover integration. Future work will focus on efficient core identification in clause sets and proof traces, on integrating CSC/S-CSC generation into saturation-based workflows as a prover-compatible lemma mechanism, and on extending the analysis to equality-rich settings such as superposition.

Acknowledgments

This research is supported by the National Natural Science Foundation of China (Grant No. 62176142), the MOE (Ministry of Education of China) Project of Humanities and Social Sciences (Grant No. 20XJCZH016), Chengdu Science and Technology Program (Grant No. 2026-RK00-00028-ZF), the Science and Technology Support Project of Sichuan Province (Grant No. 2024YFHZ0316) and the Fundamental Research Funds for the Central Universities (Grant No. 2682024ZTPY041).

References

[1] Bernhard Beckert and Reiner Hähnle. Reasoning and verification: State of the art and current trends. *IEEE Intelligent Systems*, 29(1):20–29, 2014.

[2] Feng Cao, Yang Xu, Shuwei Chen, Jian Zhong, and Guanfeng Wu. A contradiction separation dynamic deduction algorithm based on optimized proof search. *International Journal of Computational Intelligence Systems*, 12(2):1245–1254, 2019.

[3] Feng Cao, Yang Xu, Jun Liu, Shuwei Chen, and Jianbing Yi. A multi-clause dynamic deduction algorithm based on standard contradiction separation rule. *Information Sciences*, 566:281–299, 2021.

[4] Christian Fermüller and Alexander Leitsch. Hyperresolution and automated model building. *Journal of Logic and Computation*, 6(2):173–203, 1996.

[5] Xingxing He, Yingfang Li, and Yanghe Feng. On structures of regular standard contradictions in propositional logic. *Information Sciences*, 586:261–278, 2022.

[6] Xingxing He, Lan Pan, Yingfang Li, Jun Liu, and Luis Martínez. On structures of sign-boundary and diagonal vacancy-type standard contradictions. *Mathematical Logic Quarterly*, 72(2):e70010, 2026.

[7] Chang Chin Liang and Richard C Lee. *Symbolic logic and mechanical theorem proving.* Academic Press, New York, 1973.

[8] Xuhua Liu. *Automated reasoning based on resolution method.* Press of China Science, Beijing, 1994.

[9] William McCune. Release of prover9. In *Mile high conference on quasigroups, loops and nonassociative systems, Denver, Colorado*, 2005.

[10] Alexandre Riazanov and Andrei Voronkov. Vampire 1.1. In *International Joint Conference on Automated Reasoning*, pages 376–380. Springer, 2001.

[11] Alan JA Robinson and Andrei Voronkov. *Handbook of automated reasoning*, volume 1. Elsevier, 2001.

[12] John Alan Robinson. A machine-oriented logic based on the resolution principle. *Journal of the ACM (JACM)*, 12(1):23–41, 1965.

[13] Stephan Schulz. E–a brainiac theorem prover. *AI Communications*, 15(2-3):111–126, 2002.

[14] Geoff Sutcliffe and Martin Desharnais. The 11th IJCAR automated theorem proving system competition–CASC-J11. *AI Communications*, 36(2):73–91, 2023.

[15] Geoff Sutcliffe, Christian Suttner, Lars Kotthoff, C Raymond Perrault, and Zain Khalid. An empirical assessment of progress in automated theorem proving. In *International Joint Conference on Automated Reasoning*, pages 53–74. Springer, 2024.

[16] Leiming Tang and Xingxing He. The constructions of complete contradiction based on propositional logic. In *Developments of Artificial Intelligence Technologies in Computation and Robotics: Proceedings of the 14th International FLINS Conference (FLINS 2020)*, pages 276–283. World Scientific, 2020.

[17] Ming Xu. *Lecture on symbolic logic.* Wuhan University Press, Wuhan, 2008.

[18] Yang Xu, Shuwei Chen, Xiaomei Zhong, Jun Liu, and Xingxing He. Towards multi-clause automated deduction and theorem generation: Constructing and applying standard contradictions. *Knowledge-Based Systems*, 334:114985, 2026.

[19] Yang Xu, Jun Liu, Shuwei Chen, Xiaomei Zhong, and Xingxing He. Contradiction separation based dynamic multi-clause synergized automated deduction. *Information Sciences*, 462:93–113, 2018.

[20] Guoyan Zeng, Shuwei Chen, Jun Liu, Yang Xu, and Peiyao Liu. A complementary ratio based clause selection method for contradiction separation dynamic deduction. *Knowledge-Based Systems*, 284:111238, 2024.

[21] Jian Zhong, Yang Xu, and Feng Cao. A novel combinational ATP based on contradiction separation for first-order logic. *International Journal of Computational Intelligence Systems*, 13(1):672–680, 2020.

Received July 2025

Kalman–Galois Connections on Pseudocomplemented Distributive Lattices

Nohelia Paloma Ramos
CONICET and Instituto de Ciencias Básicas, Universidad Nacional de San Juan, 5400 San Juan, Argentina.
nprm96@gmail.com

Abstract

We introduce and develop the theory of Kalman–Galois connections on pseudocomplemented distributive lattices (KGp-algebras). We establish their fundamental properties, study KGp-Boolean filters and KGp-Boolean congruences, and prove a Glivenko-type theorem that relates KGp-algebras to Boolean algebras with additional operators. Using the Kalman construction, we prove a categorical equivalence between the algebraic category of KGp-algebras and the algebraic category of centered Kleene algebras with intuitionistic negation endowed with a unary operator g (gKANc-algebras). We also extend the Kalman construction to the Sendlewski construction and establish a categorical equivalence between the category of pairs consisting of a KGp-algebra and a Boolean KGp-filter and the algebraic category of gKAN-algebras. Additionally, we introduce the Monteiro construction, which allows us to associate to every gKAN-algebra a gKANc-algebra into which it is embedded. These results enable us to clarify the relationship between the Kalman and Monteiro constructions in the case of gKAN-algebras.

1 Introduction

Galois connections and adjunctions constitute fundamental tools for establishing structured relationships between mathematical objects. By preserving or inverting order-theoretic properties, they facilitate the transfer of structural insights across diverse domains. This mediating capability has rendered Galois connections indispensable in multiple disciplines, including mathematical logic, order theory, theoretical computer science, and Soft Computing [18].

The formal theory of Galois connections originates in the pioneering work of Øystein Ore [35], who established a general framework for relating ordered structures.

Subsequent developments have seen this concept play a pivotal role in characterizing algebraic systems, particularly Hilbert algebras, distributive lattices, and Heyting algebras [3, 4, 13, 15, 17, 20–22, 27, 34]. Among these, distributive pseudocomplemented lattices provide a particularly fertile framework for investigating order-theoretic logical operations and their algebraic manifestations.

Within a Heyting algebra $\langle A, \vee, \wedge, \Rightarrow, 0, 1\rangle$, a Galois connection is a pair of functions (f, g) characterized by the Galois connection condition

$$f(x) \leq y \iff x \leq g(y),$$

as studied by Dzik *et al.* [19]. A notable subclass, referred to as *Kalman–Galois connections* [8], satisfies the additional requirements $g(0) = 0$ and $g(x) \wedge f(y) \leq f(x \wedge y)$.

Considering the Kalman construction, the authors of [8] show that the algebraic category of Heyting algebras with Kalman–Galois connections is equivalent to the algebraic category of centered Nelson algebras endowed with a unary operator. The literature contains numerous equivalences of this type for various kinds of algebras using the Kalman construction (see, for example, [40, 41]). In particular, in [36], the authors consider *tense Heyting algebras* and prove that the category of these algebras is categorially equivalent to the category of *tense centered Nelson algebras*; similar results were demonstrated in [23] for other type of tense operators on Heyting algebras.

A *tense Heyting algebra* is an algebra $\langle A, G, H, F, P\rangle$, where A is a Heyting algebra, (G, P) and (H, F) are two pairs of Kalman–Galois connections that are interrelated by Dunn's axioms: $G(x \vee y) \leq G(x) \vee F(y)$ and $H(x \vee y) \leq H(x) \vee P(y)$. This definition was generalized by Aldo Figallo *et al.* to the case of *tense distributive lattices* [24]. It is worth mentioning that a tense distributive lattice can alternatively be defined as a structure $\langle A, G, H, F, P\rangle$, where (A, G, F) and (A, H, P) are positive modal distributive lattices according to Celani's and Jansana's definition [10–12].

The well-known fact that not all pseudocomplemented distributive lattices are Heyting algebras (cf. [37]) motivates our central objective: to extend the theory of Kalman–Galois connections to the more general setting of distributive pseudocomplemented lattices.

This paper is organized as follows: In Section 2, we present the preliminary concepts and results necessary for the development of this work, including definitions of distributive p-algebras, Galois connections, and KAN-algebras. Section 3 introduces the notion of Kalman-Galois connections on distributive p-algebras (KGp-algebras) and analyzes their basic properties. Several examples are provided to illustrate these structures, including non-trivial cases. Section 4 studies filters and congruences in the context of KGp-algebras, establishing relationships between Boolean filters stable under the operators f and g, congruences compatible with the algebraic structure, and gB-algebras. In Section 5, we prove a version of Glivenko's theorem for KGp-algebras, showing how the double negation acts as a homomorphism from each KGp-algebra into the Boolean algebra of its regular elements. Section 6 develops the theory of KGp-algebras associated with gKAN-algebras, showing how two different KGp-algebras can be constructed from a gKAN-algebra and proving they are isomorphic. Section 7 presents the Kalman construction for KGp-algebras and establishes a categorical equivalence between the categories of KGp-algebras and gKANc-algebras, extending similar constructions known for Heyting algebras. Section 8 studies congruences on $K(A)$, extending known results about congruences in distributive p-algebras to the context of KGp-algebras. Section 9 develops Sendlewski's construction for KGp-algebras, showing how every gKAN-algebra is isomorphic to one obtained through this construction applied to its quotient algebra. Finally, Section 10 presents Monteiro's construction for gKAN-algebras and shows that every gKAN-algebra is isomorphic to a subalgebra of a gKANc-algebra obtained through this construction.

2 Preliminaries

We recall the definitions and some basic properties of distributive p-algebras and KAN-algebras. For more details, see [1, 26].

2.1 Distributive p-algebras

Now, we introduce the class of pseudocomplemented distributive lattices, or distributive p-algebras for short ([1]). These structures arise as a natural generalization of Heyting algebras, obtained by retaining the notion of pseudocomplement while relaxing the requirement that it behaves as an intuitionistic negation. The pseudocomplement operation exhibits a rich algebraic behavior, leading to the distinction between dense elements, namely those whose pseudocomplement is 0. Dense elements form a filter that plays a central role in the structure of the algebra. In the sequel, we recall the fundamental properties of pseudocomplements and dense elements, which will be essential for our later developments.

Definition 2.1. *A pseudocomplemented distributive lattice, or distributive p-algebra for short, is an algebra $\langle A, \vee, \wedge, ^*, 0, 1\rangle$ such that $\langle A, \vee, \wedge, 0, 1\rangle$ is a bounded distributive lattice and satisfies the following property: $a \wedge b = 0 \iff a \leq b^*$.*

We denote by PDL the variety of all distributive p-algebras. Henceforth, unless otherwise stated, we will identify a distributive p-algebra $\langle A, \vee, \wedge, ^*, 0, 1\rangle$ with its underlying set A, omitting explicit mention of the operations when it is not strictly necessary for clarity.

Let $A \in \mathrm{PDL}$. For any $a \in A$, the following hold:

(a1) $a \wedge a^* = 0$,

(a2) $a^* \wedge a^{**} = 0$,

(a3) if $a \leq b$, then $b^* \leq a^*$,

(a4) $a \leq a^{**}$,

(a5) $a^* = a^{***}$,

(a6) $a \wedge b = 0 \iff a^{**} \wedge b = 0$,

(a7) $(a \wedge b)^{**} = a^{**} \wedge b^{**}$,

(a8) $(a \vee a^*)^* = 0$,

(a9) $0^* = 1$,

(a10) $(a \vee b)^* = a^* \wedge b^*$,

(a11) $(a^{**} \vee b^{**})^{**} = (a \vee b)^{**}$.

Let $A \in \mathrm{PDL}$. A subset $F \subseteq A$ is a filter if it satisfies: (f1) $1 \in F$, (f2) if $x, y \in F$, then $x \wedge y \in F$, and (f3) if $x \in F$ and $x \leq y$, then $y \in F$. An element $a \in A$ is called dense if $a^* = 0$. The set of all dense elements is denoted by $D(A)$, that is, $D(A) = \{a \in A : a^* = 0\}$.

Definition 2.2. *Let $A \in \mathrm{PDL}$.*

(i) A lattice filter $F \subseteq A$ is called a Boolean filter if it contains all dense elements, that is, for all $x \in A$ such that $x^* = 0$.

(ii) A congruence R on A is called a Boolean congruence if A/R is a Boolean algebra, or equivalently, if $a \vee a^* \in [1]_R$ for all $a \in A$.

Lemma 2.1. [26] *Let $A \in \mathrm{PDL}$ and let R be a congruence on A. The following conditions are equivalent:*

(i) R is a Boolean congruence.

(ii) $[1]_R$ *is a Boolean filter.*

Moreover, if F *is a Boolean filter, then the set*

$$\Theta(F) = \{(a,b) \in A \times A : a \wedge f = b \wedge f \text{ for some } f \in F\} \tag{1}$$

is a Boolean congruence. Furthermore, the assignments $R \mapsto [1]_R$ *and* $F \mapsto \Theta(F)$ *define an order isomorphism between the partially ordered set of Boolean congruences of* A *and the partially ordered set of Boolean filters of* A.

2.2 KAN-algebras

Now, we introduce the class of Kleene algebras with intuitionistic negation, or KAN-algebras for short [26]. This introduction is motivated by the desire to better understand the relationship between strong and intuitionistic negation in Nelson algebras. The class of Nelson algebras are characterized by containing two types of negation: a strong involutive negation, denoted by $\sim$, and an intuitionistic negation defined as $\neg x := x \to 0$. Although both are part of the structure, the intuitionistic negation is not considered a primitive operation and its role remains somewhat hidden. KAN-algebras are proposed to explicitly study both negations, treating them on an equal footing. Moreover, these new structures generalize Nelson algebras in the same way that distributive p-algebra generalize Heyting algebras.

Definition 2.3. *A Kleene algebra is an algebra* $\langle T, \wedge, \vee, \sim, 0, 1\rangle$ *of type* $(2,2,1,0,0)$ *such that* $\langle T, \wedge, \vee, 0, 1\rangle$ *is a bounded distributive lattice and* $\sim$ *is an involution (that is,* $\sim\sim x = x$ *for all* $x \in T$*), satisfying the following conditions for all* $x, y \in T$*:*

(i) $\sim (x \vee y) = \sim x \wedge \sim y$.

(ii) $x \wedge \sim x \leq y \vee \sim y$.

A Kleene algebra is said to be centered if it has a center; that is, an element c *such that* $c = \sim c$.

Definition 2.4. [26] *We define the variety* KAN *of Kleene algebras with intuitionistic negation as the variety of algebras* $\langle T, \wedge, \vee, \sim, \neg, 0, 1\rangle$ *of type* $(2,2,1,1,0,0)$ *such that* $\langle T, \wedge, \vee, \sim, 0, 1\rangle$ *is a Kleene algebra and the following identities hold for all* $x, y \in T$*:*

(N1) $\neg(x \wedge \neg(x \wedge y)) = \neg(x \wedge \neg y)$,

(N2) $\neg(x \vee y) = \neg x \wedge \neg y$,

(N3) $x\wedge \sim x = x \wedge \neg x$,

(N4) $\sim x \leq \neg x$,

(N5) $\neg(x \wedge y) = \neg((\sim \neg x) \wedge y)$.

An algebra $\langle T, \vee, \wedge, \sim, \neg, c, 0, 1\rangle$ *of type* $(2, 2, 1, 1, 0, 0, 0)$ *is a* centered KAN-algebra, *or* KANc-algebra, *if* $\langle T, \vee, \wedge, \sim, \neg, 0, 1\rangle$ *is a KAN-algebra and the structure* $\langle T, \vee, \wedge, \sim, c, 0, 1\rangle$ *is a centered Kleene algebra.*

Remark 2.1. *Although the choice of axioms in the previous definition is technically motivated, axioms (N1) and (N2) are also satisfied in Heyting algebras when considering the intuitionistic negation. Axioms (N3) through (N5) establish the relationship between the two types of negations. In particular, axiom (N4) is referred to as the* normality axiom *in symmetric Heyting algebras (see [33]).*

We denote the variety of KAN-algebras by KAN and the variety of centered KAN-algebras by KANc.

Henceforth, unless otherwise stated, we will identify a KAN-algebra $\langle T, \vee, \wedge, \sim, \neg, 0, 1\rangle$ with its underlying set T, omitting explicit mention of the operations when it is not strictly necessary for clarity.

The study of Kleene algebras has been extended through various frameworks incorporating additional operations. A significant subclass is that of *pseudocomplemented Kleene algebras* [38].

An algebra $\langle T, \vee, \wedge, \sim, ^*, 0, 1\rangle$ is called a pseudocomplemented Kleene algebra provided $\langle T, \vee, \wedge, \sim, 0, 1\rangle$ forms a Kleene algebra and $\langle T, \vee, \wedge, ^*, 0, 1\rangle$ constitutes a distributive p-algebra. It is worth noting that this definition does not impose any axioms relating the pseudocomplement * to the Kleene negation $\sim$. In response to this, several researchers have introduced algebraic structures in which these unary operations are interrelated through additional axioms, as illustrated in [6, 28, 29].

The following example from [7] shows that the intuitionistic negation $\neg$ need not coincide with the pseudocomplement * in a Kleene algebra.

Example 2.1. *Consider the three-element chain* $T = \{0, c, 1\}$ *represented by the following Hasse diagram:*

Define the unary operators $\sim$, $\neg$ *and* * *as follows:*

x	0	c	1
$\sim x$	1	c	0
$\neg x$	1	1	0
x^*	1	0	0

It can be checked that $\langle T, \vee, \wedge, \sim, \neg, c, 0, 1\rangle$ *is a centered KAN-algebra and that* $\langle T, \vee, \wedge, \sim, ^*, 0, 1\rangle$ *is a pseudocomplemented Kleene algebra.*

The following lemma lists some properties of KAN-algebras that will be needed in the sequel. Their proofs can be found in [26].

Proposition 2.1. *Let* $T \in \mathrm{KAN}$ *and* $x \in T$. *Then:*

(N6) $\neg(x \wedge \sim x) = \neg(x \wedge \neg x) = 1$,

(N7) $\sim \neg x \leq x \leq \neg \sim x$,

(N8) $\sim \neg x \leq \neg\neg x \leq \neg \sim x$,

(N9) $\neg \sim \neg x = \neg x$.

For any $T \in \mathrm{KAN}$, we adopt the following notation:

$$\Diamond x := \sim \neg x \quad \text{and} \quad \Box x := \neg \sim x.$$

Moreover, it holds that $\Box x = \sim \Diamond \sim x$ and $\Diamond x = \sim \Box \sim x$.

Proposition 2.2. [26] *Let* $T \in \mathrm{KAN}$. *Then:*

1. $\neg 1 = 0$ *and* $\neg 0 = 1$,
2. *If* $x \leq y$, *then* $\neg y \leq \neg x$,
3. $\neg(x \wedge \sim x) = \neg(x \wedge \neg x) = 1$,

4. $\Diamond 0 = 0$ *and* $\Box 1 = 1$,

5. $\Diamond 1 = 1$ *and* $\Box 0 = 0$,

6. $\Diamond x \leq x \leq \Box x$,

7. $\Diamond x \leq \neg\neg x \leq \Box x$,

8. $\neg\Diamond x = \Box\neg x = \neg x$.

Lemma 2.2. [26] *Let* $T \in \mathrm{KAN}$. *Then,* $x \leq y$ *if and only if* $\Diamond x \leq \Diamond y$ *and* $\Box x \leq \Box y$.

The *regularity property* has been investigated by several authors in various types of algebras, including ukasiewicz–Moisil algebras [5], Nelson algebras [33], and distributive double *p*-algebras [32], among others. The following lemma establishes that a similar notion of regularity also exists in the context of KAN-algebras.

Lemma 2.3. [7]*(Regularity Property) Let* $T \in \mathrm{KAN}$. *If* $\Box x = \Box y$ *and* $\Diamond x = \Diamond y$, *then* $x = y$.

In what follows, we present several results that will be used in subsequent sections.

Proposition 2.3. [7] *Let* $T \in \mathrm{KAN}$. *Then, the following hold:*

1. $\Diamond\Diamond x = \Diamond x$,

2. $\Diamond(x \vee y) = \Diamond x \vee \Diamond y$,

3. $\Diamond(\Diamond x \wedge \Diamond y) = \Diamond(x \wedge y)$,

4. $\Box\Box x = \Box x$,

5. $\Box(x \wedge y) = \Box x \wedge \Box y$,

6. $\Box(\Box x \vee \Box y) = \Box(x \vee y)$,

7. $x \vee \sim x = \sim x \vee \Diamond x$,

8. $x \wedge \sim x = x \wedge \Box \sim x$,

9. $x = (\Diamond x \vee \sim x) \wedge \Box x$.

Lemma 2.4. [7] *Let* $T \in \mathrm{KANc}$. *Then:*

1. $\neg c = 1$,

2. $\Diamond c = 0$ *and* $\Box c = 1$,

3. $\Diamond(c \wedge \Box x) = 0$ *and* $\Box(\Diamond x \vee c) = 1$,

4. $x = (\Diamond x \vee c) \wedge \Box x$,

5. $x \vee c = \Diamond x \vee c$.

Lemma 2.5. [7] *Let* $T \in \mathrm{KAN}$ *and* $c \in T$. *Then,* c *is the center of* T *if and only if* $\Diamond c = 0$ *and* $\Box c = 1$.

Lemma 2.6. [26] *Let* $T \in \mathrm{KAN}$, *and let* $\theta \subseteq T \times T$ *be the relation defined as*

$$(x, y) \in \theta \text{ if and only if } \neg x = \neg y. \tag{2}$$

Then, θ *is an equivalence relation and is a congruence with respect to* $\wedge$, $\vee$, *and* $\neg$.

For each $T \in \mathrm{KAN}$, with a slight abuse of notation, $\langle T_\theta, \wedge, \vee, \neg, 0, 1\rangle$ will denote a new algebraic structure whose operations $\wedge$, $\vee$, and $\neg$ are well defined due to Lemma 2.6.

Lemma 2.7. [26] *If* $T \in \mathrm{KAN}$, *then* $T/\theta \in \mathrm{PDL}$.

3 Kalman-Galois Connections

In this section, we introduce the notion of Kalman-Galois connections on distributive p-algebras. We provide several examples to illustrate the motivation behind studying these structures. Furthermore, we analyze some basic properties that will be relevant for their characterization in subsequent sections.

Definition 3.1. *A distributive p-algebra with a Kalman-Galois connection, or simply a KGp-algebra, is a structure* $\langle A, f, g\rangle$, *where* $A \in \mathrm{PDL}$ *and* (f, g) *is a Galois connection on* A *such that:*

(G1) $g(0) = 0$,

(G2) $g(x) \wedge f(y) \leq f(x \wedge y)$,

(G3) $g(x \vee y) \leq g(x) \vee f(y)$,

(G4) $g(x)^* \leq f(x^*)$,

(G5) $f(x)^* \leq g(x^*)$.

Remark 3.1. *The chosen axiomatization is justified by both technical and methodological considerations. Axiom (G3) enables the Sendlewski construction while preserving the natural notion of KGp-Boolean filter, while axioms (G4) and (G5) provide the necessary foundation for proving a Glivenko-type theorem in this generalized setting. This expanded axiomatic system reflects the fact that pseudocomplemented lattices form a strictly wider class than Heyting algebras, requiring additional constraints to maintain the intended behavior of the Kalman-Galois connection. While a weaker definition could be adopted, this would come at the cost of a less powerful theory with fewer available results. The present axiomatization thus achieves an optimal balance between generality and demonstrative power, supporting all central constructions of this work in a uniform framework.*

Next, we present some properties of KGp-algebras.

Lemma 3.1. *Let $\langle A, f, g\rangle$ be a KGp-algebra. Then:*

(G6) $f(0) = 0$ *and* $g(1) = 1$,

(G7) $f(x \vee y) = f(x) \vee f(y)$,

(G8) $g(x \wedge y) = g(x) \wedge g(y)$,

(G9) $x \leq gf(x)$,

(G10) $fg(x) \leq x$,

(G11) *If* $x \leq y$, *then* $g(x) \leq g(y)$,

(G12) *If* $x \leq y$, *then* $f(x) \leq f(y)$,

(G13) $f(x^*) = g(x)^*$,

(G14) $g(x^*) = f(x)^*$,

(G15) $gfg(x) = g(x)$ *and* $fgf(x) = f(x)$,

(G16) $g(x) \leq f(x)$.

Proof. Since (f, g) forms a Galois connection, properties (G6)(G12) and (G15) are immediate consequences of the definition of a Galois connection. We shall now prove (G13). Let $x \in A$. By (G2) and (G6), we have $g(x) \wedge f(x^*) \leq f(x \wedge x^*) = f(0) = 0$. Hence, $f(x^*) \leq g(x)^*$. Finally, applying (G4), we conclude that $f(x^*) = g(x)^*$. Property (G14) can be proved similarly. Moreover, by (G3), we have $g(0 \vee x) \leq g(0) \vee f(x)$, and then, by (G1), it follows that $g(x) \leq f(x)$, which establishes (G16). □

Example 3.1. *The following example is taken from [19]. Let $A = \{0, a, 1\}$ be the three-element chain with the following order:*

The pseudocomplement table for this chain is:

x	0	a	1
x^*	1	0	0

Define the unary operators f and g on A as follows:

x	0	a	1
$f(x)$	0	1	1
$g(x)$	0	0	1

It can be verified that (f, g) forms a Galois connection on A, and that $\langle A, f, g\rangle$ is not a KGp-algebra.

Example 3.2. *Let $A \in$ PDL. A trivial example in which $\langle A, f, g\rangle$ forms a KGp-algebra is obtained by taking $f = g = \mathrm{id}_A$.*

We now present two examples that will be used later.

Example 3.3. *Let A be the lattice shown in Figure 1, where the operations *, f, and g are given in Table 1.*

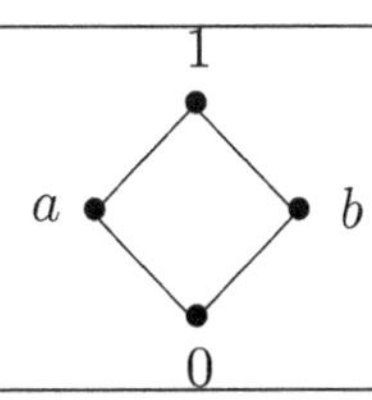

Figure 1

x	0	a	b	1
x^*	1	b	a	0
$f(x)$	0	b	1	1
$g(x)$	0	0	a	1

Table 1

Then, $\langle A, f, g\rangle$ is a KGp-algebra.

Remark 3.2. *It can be shown that if $\langle A, f, g\rangle$ is a KGp-algebra, then the following property holds:*

$$\text{If} \quad x \wedge f(y) = 0 \quad \text{then} \quad y \wedge g(x) = 0.$$

This implication follows from the fact that (f, g) forms a Galois connection on A, together with axiom (G1). *The converse does not hold in general, as shown in Example 3.3: $1 \wedge g(a) = 0$, but $a \wedge f(1) = a \neq 0$, which provides a counterexample.*

Example 3.4. *Let A be the lattice shown in Figure 2, where the operations *, f, and g are given in Table 2.*

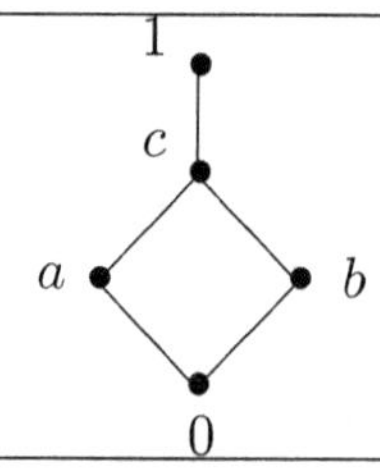

Figure 2

x	0	a	b	c	1
x^*	1	b	a	0	0
$f(x)$	0	b	a	c	1
$g(x)$	0	b	a	c	1

Table 2

Then, $\langle A, f, g\rangle$ is a KGp-algebra.

Example 3.5. *Consider the lattice $A = \mathbb{N}^2 \cup \{\bot, \top\}$, consisting of the lattice $\mathbb{N}^2$ (defined coordinatewise) with added top $\top$ and bottom $\bot$ (here $\mathbb{N}$ is the set of natural numbers). This structure forms a pseudocomplemented distributive lattice by defining*

$\neg\top = \bot$, $\neg\bot = \top$ and $\neg(m,n) = \bot$ for any $m,n \in \mathbb{N}$. However, a Heyting implication cannot be defined (see [26]). If we define f and g in A as in Example 3.2, we obtain a KGp-algebra $\langle A, f, g\rangle$ that is not a Heyting algebra with Kalman-Galois connections.

The following example illustrates a non-trivial KGp-algebra.

Example 3.6. *Let $A = \{0, a, b, c, d, e, u, v, 1\}$ be the lattice shown in Figure 3, where the operations *, f, and g are given in Table 3.*

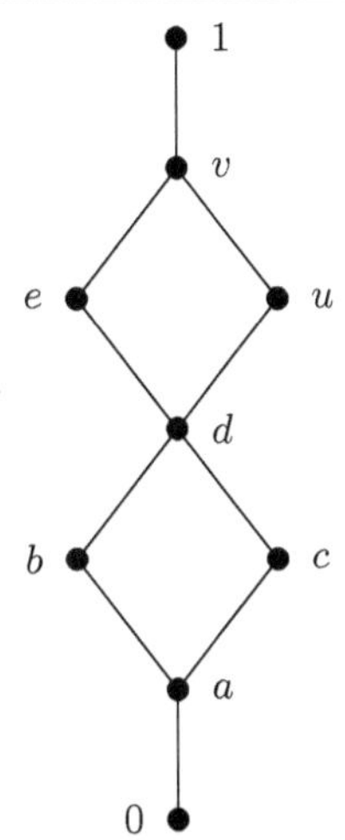

Figure 3

x	0	a	b	c	d	e	u	v	1
x^*	1	0	0	0	0	0	0	0	0
$g(x)$	0	a	c	a	c	c	c	v	1
$f(x)$	0	a	v	b	v	v	v	v	1

Table 3

It can be verified that $\langle A, f, g\rangle$ is a KGp-algebra.

Example 3.7. *A* monadic pseudocomplemented distributive lattice *(see [9]) is a structure $\langle A, \forall, \exists\rangle$, where A is a pseudocomplemented distributive lattice and $\forall$ and $\exists$ are unary operations on A satisfying the following conditions:*

(m1) $\exists 0 = 0$

(m2) $x \leq \exists x$

(m3) $\exists(x \wedge y) = \exists x \wedge \exists y$

(m4) $\exists(x \vee y) = \exists x \vee \exists y$

(m5) $\forall 1 = 1$

(m6) $\forall x \leq x$

(m7) $\forall(x \vee y) = \forall x \vee \forall y$

(m8) $\forall(x \wedge y) = \forall x \wedge \forall y$

(m9) $\forall \exists x = \exists x$

(m10) $\exists \forall x = \forall x$

(m11) $(\forall x)^* \leq \exists x^*$

(m12) $(\exists x)^* \leq \forall x^*$

(m13) $\forall(x \vee y) \leq \exists x \vee \forall y$

(m14) $\exists x \wedge \forall y \leq \exists(x \wedge y)$

If we define $g := \forall$ and $f := \exists$, then (f, g) is a Kalman–Galois connection on A.

4 Filters and Congruences

In this section, we study three fundamental notions in the context of KGp-algebras: Boolean filters closed under the operator g, congruences compatible with the algebraic structure, and gB-algebras. We analyze the conditions that relate these concepts and their role in the general theory of KGp-algebras.

Definition 4.1. *Let $\langle A, f, g\rangle$ be a KGp-algebra. A Boolean filter F of A is called a KGp-Boolean filter if it satisfies the following condition:*

$$x \in F \text{ then } g(x) \in F.$$

Remark 4.1. *By axiom (G16), every KGp-Boolean filter F is closed under f.*

Example 4.1. *Let $\langle A, f, g\rangle$ be the KGp-algebra from Example 3.4. Then, it can be verified that the set $S := \{c, 1\}$ is a KGp-Boolean filter.*

Proposition 4.1. *Let $\langle A, f, g\rangle$ be a KGp-algebra. Then, the set $D(A)$, consisting of the dense elements of A, is the smallest KGp-Boolean filter of A.*

Proof. It is known that $D(A) \subseteq A$ is a Boolean filter and the smallest one with respect to inclusion. We now show that it is closed under g. Let $x \in D(A)$. Then, $x^* = 0$. By axioms (G13) and (G6), we have $g(x)^* = f(x^*) = f(0) = 0$. Hence $g(x) \in D(A)$, as required. □

While Galois connections on Boolean lattices have been extensively studied in the literature [30, 39], we focus here on a particular subclass of Galois connections defined on Boolean algebras.

We now introduce the definition of a gB-algebra, which provides the appropriate framework to establish the equivalence between the partially ordered set of Boolean congruences of a KGp-algebra and the partially ordered set of its Boolean filters. This notion will also serve as a fundamental tool in the formulation and proof of Glivenko's theorem for these structures.

Definition 4.2. *An algebra $\langle B, f, g\rangle$ is called a* gB-algebra *if $\langle B, \vee, \wedge, \neg, 0, 1\rangle$ is a Boolean algebra, and the unary operators f and g satisfy the following conditions:*

(B1) $g(0) = 0$,

(B2) $g(x \wedge y) = g(x) \wedge g(y)$,

(B3) $x \leq gf(x)$,

(B4) $g(x) = \neg f(\neg x)$,

(B5) $g(x) \wedge f(y) \leq f(x \wedge y)$.

The following lemma collects some basic properties of gB-algebras. Their proofs follow arguments analogous to those used for KGp-algebras, and are therefore omitted.

Lemma 4.1. *Let $\langle B, f, g\rangle$ be a gB-algebra. Then:*

(B6) $f(0) = 0$ *and* $g(1) = 1$,

(B7) $f(x \vee y) = f(x) \vee f(y)$,

(B8) $g(x \vee y) \leq g(x) \vee f(y)$,

(B9) $fg(x) \leq x$,

(B10) If $x \leq y$, *then* $g(x) \leq g(y)$,

(B11) If $x \leq y$, then $f(x) \leq f(y)$,

(B12) $f(\neg x) = \neg g(x)$,

(B13) $g(\neg x) = \neg f(x)$,

(B14) $gfg(x) = g(x)$ and $fgf(x) = f(x)$,

(B15) $g(x) \leq f(x)$.

Remark 4.2. *From Lemma 4.1, it follows that if $\langle B, f, g\rangle$ is a gB-algebra, then it is also a KGp-algebra.*

Definition 4.3. *Let $\langle A, f, g\rangle$ be a KGp-algebra. A Boolean congruence ψ is called a KGp-Boolean congruence of A if the quotient algebra $\langle A/\psi, f_\psi, g_\psi\rangle$ is a gB-algebra, where the operations f_ψ and g_ψ are defined by $f_\psi([x]_\psi) = [f(x)]_\psi$ and $g_\psi([x]_\psi) = [g(x)]_\psi$.*

Lemma 4.2. *Let $\langle A, f, g\rangle$ be a KGp-algebra. The following conditions are equivalent:*

(i) ψ is a KGp-Boolean congruence of A,

(ii) ψ is a Boolean congruence of A which is compatible whith f and g.

Proof. (i)⇒(ii): Suppose $(x, y) \in \psi$. Then, $[x]_\psi = [y]_\psi$, and by the definition of f_ψ it follows that $[f(x)]_\psi = [f(y)]_\psi$, hence $(f(x), f(y)) \in \psi$. An analogous argument shows that ψ is also preserved by g.

(ii)⇒(i): Conditions (B1), (B2), (B3), and (B5) are inherited from (G1), (G8), (G9), and (G2), respectively. Furthermore, by (G13) we have $f(x^*) = g(x)^*$, and therefore $[f(x^*)]_\psi = [g(x)^*]_\psi$. By hypothesis, $f_\psi([x]_\psi^*) = f_\psi([x^*]_\psi) = g_\psi([x]_\psi)^*$. Taking pseudocomplements in the quotient yields $f_\psi([x]_\psi^*)^* = g_\psi([x]_\psi)$, which establishes (B4). Thus $\langle A/\psi, f_\psi, g_\psi\rangle$ is a gB-algebra. □

Lemma 4.3. *Let $\langle A, f, g\rangle$ be a KGp-algebra. If ψ is a KGp-Boolean congruence, then the class $[1]_\psi$ is a KGp-Boolean filter of A.*

Proof. By Lemma 2.1, $[1]_\psi$ is a Boolean filter. Since $x \in [1]_\psi$, we have $(x, 1) \in \psi$, and Lemma 4.2 ensures that $(g(x), g(1)) \in \psi$. Using property (G6), it follows that $g(1) = 1$, so $g(x) \in [1]_\psi$. Thus, $[1]_\psi$ is closed under g, and therefore it is a KGp-Boolean filter.

□

Lemma 4.4. *Let $\langle A, f, g\rangle$ be a KGp-algebra and S a KGp-Boolean filter. Then, the set $\Theta(S)$ defined in* (1) *is a KGp-Boolean congruence on A.*

Proof. By Lemma 2.1, the relation $\Theta(S)$ is a Boolean congruence on A. We now prove that $\Theta(S)$ is closed under f and g. Suppose that $(a, b) \in \Theta(S)$, so there exists $s \in S$ such that $a \wedge s = b \wedge s$. By uniqueness of g, we have $g(a \wedge s) = g(b \wedge s)$, and by (G8), it follows that $g(a) \wedge g(s) = g(b) \wedge g(s)$. Since S is a KGp-Boolean filter, we know that $g(s) \in S$, which implies that $(g(a), g(b)) \in \Theta(S)$.

Now, using (G2) and (G12), we obtain $f(a) \wedge g(s) \leq f(a \wedge s) = f(b \wedge s) \leq f(b)$, so $f(a) \wedge g(s) \leq f(b) \wedge g(s)$. Similarly, $f(b) \wedge g(s) \leq f(a) \wedge g(s)$, and hence $f(a) \wedge g(s) = f(b) \wedge g(s)$, which shows that $(f(a), f(b)) \in \Theta(S)$. Therefore, by Lemma 4.2, the relation $\Theta(S)$ is a KGp-Boolean congruence. □

The following result follows from Lemmas 4.3 and 4.4.

Theorem 4.2. *Let $\langle A, f, g\rangle$ be a KGp-algebra. Then, the assignments*

$$\psi \mapsto [1]_\psi \quad \textit{and} \quad S \mapsto \Theta(S)$$

define an order isomorphism between the partially ordered set of KGp-Boolean congruences of A and the partially ordered set of KGp-Boolean filters of A.

5 Glivenko's Theorem

In [25], V. Glivenko proved that a proposition is classically provable if and only if its double negation is intuitionistically provable. This result admits an algebraic formulation: the double negation acts as a homomorphism from each Heyting algebra into the Boolean algebra of its regular elements. There exist various versions of this theorem, both in logical and algebraic form, adapted to other logical systems and to different classes of algebras (for further details, see [2, 16]). In this section, we establish a version of this theorem for KGp-algebras.

Let $A \in$ PDL. An element $a \in A$ is called regular if $a = a^{**}$. We define the set of regular elements as

$$R(A) := \{a \in A : a^{**} = a\}.$$

As stated in [1], if we define the operations

$$a \vee_r b := (a \vee b)^{**}, \quad a \wedge_r b := a \wedge b, \quad \neg_r a := a^*, \quad 0_r := 0, \quad 1_r := 1,$$

then the structure $\langle R(A), \vee_r, \wedge_r, \neg_r, 0_r, 1_r\rangle$ forms a Boolean algebra. Moreover, it can be shown that $a \in R(A)$ if and only if there exists $b \in A$ such that $a = b^*$.

Given a KGp-algebra $\langle A, f, g\rangle$, our next goal is to associate to it a gB-algebra. For this purpose, we consider the following unary operations defined on $R(A)$:

$$g_r(a) := g(a)^{**} \quad \text{and} \quad f_r(a) := f(a)^{**}.$$

Lemma 5.1. *Let $\langle A, f, g\rangle$ be a KGp-algebra and let $a \in R(A)$. Then, the following hold:*

(i) $g_r(a) \in R(A)$,

(ii) $f_r(a) = \neg_r g_r(\neg_r a)$,

(iii) $f_r(a) \in R(A)$.

Proof. We prove only item (ii). Let $a \in R(A)$. Using properties (G14) and (a5), we compute:

$$\neg_r g_r(\neg_r a) = \neg_r g_r(a^*) = g(a^*)^{***} = f(a)^{**} = f_r(a).$$

□

Theorem 5.1. *Let $\langle A, f, g\rangle$ be a KGp-algebra. Then, $\langle R(A), f_r, g_r\rangle$ is a gB-algebra.*

Proof. It is known that $R(A)$ is a Boolean algebra and, by Lemma 5.1, the operators f_r and g_r are well-defined. Therefore, it suffices to verify axioms (B1)(B5). Let $x, y \in R(A)$.
(B1): By (G1), we have $g_r(0) = g(0)^{**} = 0^{**} = 1^* = 0$.
(B2): Using (G8), we get $g_r(x \wedge y) = g(x \wedge y)^{**} = (g(x) \wedge g(y))^{**}$. Moreover, by (a7), $(g(x) \wedge g(y))^{**} = g(x)^{**} \wedge g(y)^{**}$. Hence, $g_r(x \wedge y) = g_r(x) \wedge g_r(y)$.
(B3): Applying (G9), (a3), and (G13), we deduce that

$$x^{****} \leq gf(x)^{****} = f(f(x)^*)^{***} = g(f(x)^{**})^{**} = g_r f_r(x).$$

Since $x \in R(A)$, it follows that $x \leq g_r f_r(x)$.
(B4): By (G13) and the definition of g_r, we have

$$\neg_r f_r(\neg_r x) = f(x^*)^{***} = g(x)^{****} = g_r(x)^{**} = g_r(x).$$

(B5): Using axioms (G2) and (a7), it follows that

$$g_r(x) \wedge f_r(y) = g(x)^{**} \wedge f(y)^{**} = (g(x) \wedge f(y))^{**} \leq f(x \wedge y)^{**} = f_r(x \wedge y).$$

Therefore, we conclude that $\langle R(A), f_r, g_r\rangle$ is a gB-algebra. □

Example 5.1. *Consider the KGp-algebra from Example 3.4. Its regular elements are $R(A) = \{0, a, b, 1\}$, depicted in Figure 4. With the operations f_r and g_r given in Table 4, we observe that $\langle R(A), f_r, g_r\rangle$ is a gB-algebra.*

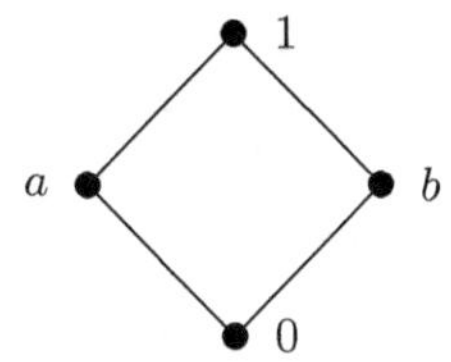

Figure 4

x	$f_r(x)$	$g_r(x)$
0	0	0
a	b	b
b	a	a
1	1	1

Table 4

For the remainder of this section, we adopt the simplified notation $\theta := \Theta(D(A))$.

Lemma 5.2. *Let $\langle A, f, g\rangle$ be a KGp-algebra. Then, $\langle A/\theta, f_\theta, g_\theta\rangle$ is a gB-algebra.*

Proof. By Proposition 4.1, $D(A)$ is a Boolean KGp-filter. Hence, by Lemma 4.4, θ is a Boolean KGp-congruence. Consequently, $\langle A/\theta, f_\theta, g_\theta\rangle$ forms a gB-algebra. □

Example 5.2. *Consider the KGp-algebra from Example 3.4. The set of dense elements is given by $D(A) = \{c, 1\}$. The associated Boolean KGp-congruence is:*

$$\theta = \{(0,0),\ (a,a),\ (b,b),\ (c,c),\ (1,1),\ (c,1),\ (1,c)\}.$$

The quotient algebra is $A/\theta = \{[0], [a], [b], [1]\}$, where $[0] = \{0\}, [a] = \{a\}, [b] = \{b\}, [1] = \{c, 1\}$. It can be verified that $\langle A/\theta, f_\theta, g_\theta\rangle$ is a gB-algebra, where the Boolean structure is depicted in Figure 5 and the operators f_θ and g_θ are defined as in Table 5.

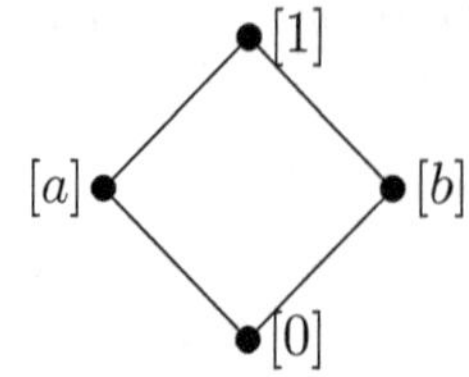

Figure 5

$[x]$	$f_\theta([x])$	$g_\theta([x])$
$[0]$	$[0]$	$[0]$
$[a]$	$[b]$	$[b]$
$[b]$	$[a]$	$[a]$
$[1]$	$[1]$	$[1]$

Table 5

The following theorem represents a version of Glivenko's theorem, which constitutes the main object of study in this section.

Theorem 5.2. *Let $\langle A, f, g\rangle$ be a KGp-algebra and $\langle R(A), f_r, g_r\rangle$ the associated gB-algebra. Then, the map $r_A : A \to R(A)$, defined by $r_A(x) = x^{**}$ for all $x \in A$, is a surjective homomorphism.*

Proof. It is known that r_A is a surjective homomorphism; it remains to prove that r_A commutes with f and g. Using the definitions of r_A and g_r, together with properties (G13) and (G14) and axiom (a5), we have

$$\begin{aligned} g_r(r_A(x)) &= g_r(x^{**}) \\ &= g(x^{**})^{**} \\ &= f(x^*)^{***} \\ &= f(x^*)^* \\ &= g(x)^{**} \\ &= r_A(g(x)). \end{aligned}$$

The case for f is analogous, and therefore r_A commutes with both f and g.

□

Corollary 5.3. *Let $\langle A, f, g\rangle$ be a KGp-algebra. Then, the gB-algebras $\langle R(A), f_r, g_r\rangle$ and $\langle A/\theta, f_\theta, g_\theta\rangle$ are isomorphic.*

Proof. By Theorem 5.2, the function $r_A : A \to R(A)$ is a surjective homomorphism. Moreover, it can be shown that θ is the kernel of r_A. Therefore, by the First Isomorphism Theorem, there exists an isomorphism t between the gB-algebras $\langle R(A), f_r, g_r\rangle$ and $\langle A/\theta, f_\theta, g_\theta\rangle$, as illustrated in Figure 6, where q denotes the canonical projection.

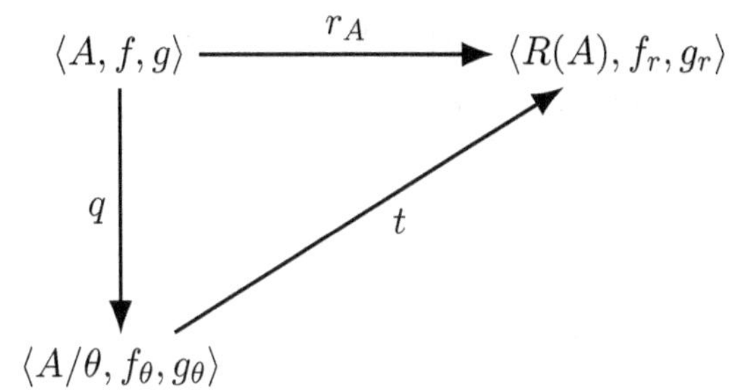

Figure 6

□

Example 5.3. *As can be observed, the algebra in Example 5.1 is isomorphic to that in Example 5.2.*

6 KGp-algebras associated with a gKAN-algebra

In this section, we introduce the notion of a gKAN-algebra and show how, from a gKAN-algebra, two different KGp-algebras can be constructed, which turn out to be isomorphic.

Let $\langle T, \vee, \wedge, \sim, \neg, 0, 1\rangle$ be a KAN-algebra, and let g be a unary operator on T. We define the operator $f(x) :=\sim g(\sim x)$.

Definition 6.1. *An algebra $\langle T, g\rangle$ is called a gKAN-algebra if $\langle T, \vee, \wedge, \sim, \neg, 0, 1\rangle$ is a KAN-algebra and g is a unary operation on T satisfying the following conditions:*

(K1) $g(0) = 0$,

(K2) $g(x \wedge y) = g(x) \wedge g(y)$,

(K3) $x \leq gf(x)$,

(K4) $\neg g(x) = f(\neg x)$,

(K5) $\neg f(x) = g(\neg x)$,

(K6) $g(x) \wedge f(y) \leq f(x \wedge y)$.

Moreover, we say that it is a centered gKAN-algebra, or gKANc-algebra, if $\langle T, \vee, \wedge, \sim, \neg, c, 0, 1\rangle$ *is a KANc-algebra and* $g(c) = c$.

Next, we present some properties of gKAN-algebras.

Proposition 6.1. *Let* $\langle T, g\rangle$ *be a gKAN-algebra. Then:*

(K7) $g(1) = 1$,

(K8) $f(0) = 0$ *and* $f(1) = 1$,

(K9) $f(x \vee y) = f(x) \vee f(y)$,

(K10) $fg(x) \leq x$,

(K11) If $x \leq y$, *then* $g(x) \leq g(y)$,

(K12) If $x \leq y$, *then* $f(x) \leq f(y)$,

(K13) $f(x) \leq y$ *if and only if* $x \leq g(y)$,

(K14) $g(x \vee y) \leq g(x) \vee f(y)$,

(K15) $\Box g(x) = g(\Box x)$,

(K16) $\Box f(x) = f(\Box x)$,

(K17) $\Diamond g(x) = g(\Diamond x)$,

(K18) $\Diamond f(x) = f(\Diamond x)$,

(K19) $\Diamond g(x) = \Diamond g(\Diamond x)$,

(K20) $\Diamond f(x) = \Diamond f(\Diamond x)$,

(K21) $\Box g(x) = \Box g(\Box x)$,

(K22) $\Box f(x) = \Box f(\Box x)$,

(K23) $\Diamond f(x) = \sim \Box g(\sim x)$,

(K24) $\Box f(x) = \sim \Diamond g(\sim x)$.

Proof. We prove only (K13), (K14), (K15), (K17), (K19), and (K21).
(K13): Let $x, y \in T$. Suppose $f(x) \leq y$. Then, by (K11) and (K3), we get $x \leq gf(x) \leq g(y)$. Conversely, suppose $x \leq g(y)$. Then, by (K12) and (K9), we have $f(x) \leq fg(y) \leq y$.
(K14): For $x, y \in T$, by (K6) we have $g(\sim x) \wedge f(\sim y) \leq f(\sim x \wedge \sim y)$. Hence, $g(x \vee y) = \sim f(\sim x \wedge \sim y) \leq \sim (g(\sim y) \wedge f(\sim x)) = \sim (\sim f(y) \wedge \sim g(x)) = g(x) \vee f(y)$.
(K15): Using (K5), we have $\neg \sim g(x) = \neg f(\sim x) = g(\neg \sim x)$, that is, $\Box g(x) = g(\Box x)$.
(K17): By (K4), it follows that $\Diamond g(x) = \sim f(\neg x) = g(\Diamond x)$.
(K19): Applying (K17), we get $\Diamond g(\Diamond x) = \Diamond\Diamond g(x) = \Diamond g(x)$.
(K21): By (K15), $\Box g(\Box x) = \Box\Box g(x) = \Box g(x)$. □

Lemma 6.1. *Let $\langle T, g \rangle$ be a gKANc-algebra. Then, the following properties hold:*

(i) $f(c) = c$,

(ii) $g(x \vee c) = g(x) \vee c$,

(iii) $f(x \wedge c) = f(x) \wedge c$.

Proof.
(i): We have $f(c) = \sim g(\sim c) = \sim g(c) = \sim c = c$.
(ii): By (K14) and item (i), we obtain $g(x \vee c) \leq g(x) \vee f(c) = g(x) \vee c$. Hence, $g(x \vee c) = g(x) \vee c$.
(iii): By (K6), it follows that $f(x \wedge c) \geq f(x) \wedge g(c) = f(x) \wedge c$, and therefore we obtain $f(x \wedge c) = f(x) \wedge c$. □

Example 6.1. *Consider the lattice T given by the Hasse diagram in Figure 7 and whose operators $\sim$, $\neg$, and g are defined in Table 6.*

x	0	a	b	c	d	e	h	1
$\sim x$	1	h	e	d	c	b	a	0
$\neg x$	1	1	1	1	c	b	a	0
$g(x)$	0	0	a	c	d	d	e	1

Table 6

Then, $\langle T, g \rangle$ is a gKAN-algebra.

Example 6.2. *A* monadic KAN-algebra *is a structure $\langle T, \exists \rangle$ (see [9]), where T is a KAN-algebra, and $\exists$ are unary operations on T verifying the following identities:*

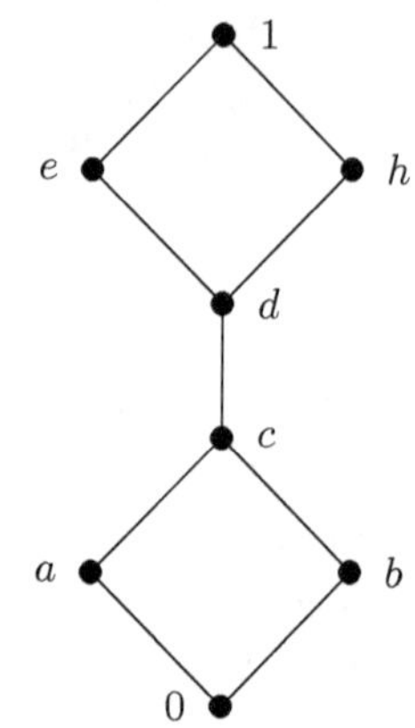

Figure 7

(M1) $\exists 0 = 0$,

(M2) $x \leq \exists x$,

(M3) $\exists(x \wedge \exists y) = \exists x \wedge \exists y$,

(M4) $\exists(x \vee y) = \exists x \vee \exists y$,

(M5) $\forall \exists x = \exists x$,

(M6) $\forall x \wedge \exists y \leq \exists(x \wedge y)$,

(M7) $\forall \neg x = \neg \exists x$,

(M8) $\neg \forall x = \exists \neg x$, *where* $\forall x := \sim \exists \sim x$.

Let $g := \forall$ *and* $f := \exists$. *Then we can prove that* $\langle T, g\rangle$ *is a gKAN-algebra.*

Let $T \in$ KAN. Consider the set

$$T^{\Diamond} := \{x \in T : \Diamond x = x\}$$

and define the following operations on $T^{\Diamond}$:

$$x \vee^{\Diamond} y := \Diamond(x \vee y),$$
$$x \wedge^{\Diamond} y := \Diamond(x \wedge y),$$
$$\neg^{\Diamond} x := \Diamond(\neg x).$$

Note that if $x \in T^{\lozenge}$, then $\sim x = \neg x$, hence $T^{\lozenge} = \{x \in T : \sim x = \neg x\}$. Moreover, if we consider the set $T^{\square} := \{x \in T : \square x = x\}$, it follows easily that $T^{\square} = \sim (T^{\lozenge})$.

Theorem 6.2. *Let $\langle T, g\rangle$ be a gKAN-algebra. Then, $\langle T^{\lozenge}, f^{\lozenge}, g^{\lozenge}\rangle$ is a KGp-algebra where the operations $f^{\lozenge}$ and $g^{\lozenge}$ are defined by:*

(i) $f^{\lozenge}(x) = \lozenge f(x)$,

(ii) $g^{\lozenge}(x) = \lozenge g(x)$.

Proof. By Proposition 2.3 (1) and (4), $f^{\lozenge}$ and $g^{\lozenge}$ are well defined. Moreover, by [7, Theorem 3.1], $T^{\lozenge}$ is a distributive p-algebra. It remains to prove that the pair $(f^{\lozenge}, g^{\lozenge})$ forms a Galois connection on $T^{\lozenge}$ and satisfies (G1)-(G5). Indeed, let $x, y \in T^{\lozenge}$ such that $f^{\lozenge}(x) \leq y$. By (K10), we have $g(f^{\lozenge}(x)) \leq g(y)$. Then, by Lemma 2.2, we get $g^{\lozenge}(f^{\lozenge}(x)) \leq g^{\lozenge}(y)$. Also, by (K19), $g^{\lozenge}(f^{\lozenge}(x)) = \lozenge g(\lozenge f(x)) = \lozenge g(f(x))$. Then, by (K3) and Lemma 2.2, it follows that $x = \lozenge x \leq \lozenge g(f(x)) \leq g^{\lozenge}(y)$. Similarly, if $x \leq g^{\lozenge}(y)$ then $f^{\lozenge}(x) \leq y$. Therefore, $(f^{\lozenge}, g^{\lozenge})$ forms a Galois connection.
(G1): By (K1) and Proposition 2.2, $g^{\lozenge}(0) = \lozenge g(0) = \lozenge 0 = 0$.
(G2): Using Proposition 2.3 (3), (K6), and (K20), we have

$$\begin{aligned} g^{\lozenge}(x) \wedge^{\lozenge} f^{\lozenge}(y) &= \lozenge(\lozenge g(x) \wedge \lozenge f(y)) \\ &= \lozenge(g(x) \wedge f(y)) \\ &\leq \lozenge f(x \wedge y) \\ &= \lozenge f(\lozenge(x \wedge y)) \\ &= f^{\lozenge}(x \wedge^{\lozenge} y). \end{aligned}$$

(G3): Using Proposition 2.3 (3), (K14), and (K19), we have

$$\begin{aligned} g^{\lozenge}(x) \vee^{\lozenge} f^{\lozenge}(y) &= \lozenge(\lozenge g(x) \vee \lozenge f(y)) \\ &= \lozenge(g(x) \vee f(y)) \\ &\geq \lozenge g(x \vee y) \\ &= \lozenge g(\lozenge(x \vee y)) \\ &= g^{\lozenge}(x \vee^{\lozenge} y). \end{aligned}$$

(G4): Using Proposition 2.2 (8), (K4), and (K20), we obtain

$$\begin{aligned} \neg^{\lozenge} g^{\lozenge}(x) &= \lozenge(\neg \lozenge g(x)) \\ &= \lozenge(\neg g(x)) \\ &= \lozenge f(\neg x) \\ &= \lozenge f(\lozenge \neg x) \\ &= f^{\lozenge}(\neg^{\lozenge} x). \end{aligned}$$

(G5): Using (K18) and (K5), we have

$$\begin{aligned}\neg^{\Diamond} f^{\Diamond}(x) &= \Diamond(\neg\Diamond f(x))\\ &= \Diamond g(\neg\Diamond x)\\ &= g^{\Diamond}(\neg^{\Diamond} x).\end{aligned}$$

Hence, $\langle T^{\Diamond}, f^{\Diamond}, g^{\Diamond}\rangle$ is a KGp-algebra.

□

Example 6.3. *Consider the KAN-algebra from Example 6.1. The elements of $T^{\Diamond}$ are depicted in Figure 8 and the operations $\neg^{\Diamond}$, $f^{\Diamond}$, and $g^{\Diamond}$ are defined in Table 7.*

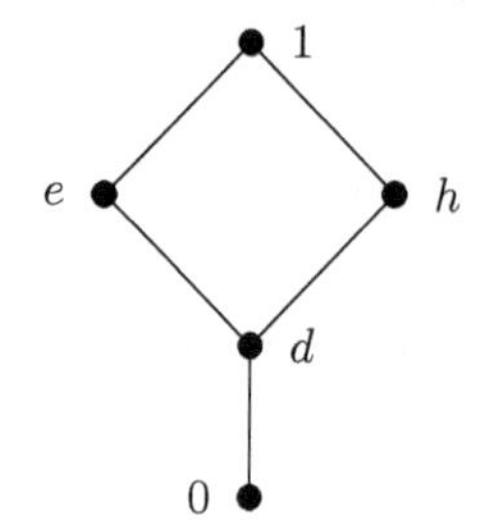

Figure 8

x	$f^{\Diamond}(x)$	$g^{\Diamond}(x)$	$\neg^{\Diamond} x$
0	0	0	1
d	d	d	0
e	h	d	0
h	1	e	0
1	1	1	0

Table 7

There exists a construction that connects distributive p-algebras with KAN-algebras. In [26], the authors showed that if T is a KAN-algebra, then the binary relation $\theta \subseteq T \times T$ defined in (2) is an equivalence relation compatible with the operations $\vee$, $\wedge$, and $\neg$. We denote by $[x]_\theta$ the equivalence class of x modulo θ, and by $T_\theta = \{[x]_\theta : x \in T\}$ the set of all equivalence classes. In this way, the structure $\langle T_\theta, \vee_\theta, \wedge_\theta, \neg_\theta, [0]_\theta, [1]_\theta\rangle$ is a distributive p-algebra, where the operations on T_θ are defined as follows:

$$
\begin{aligned}
[x]_\theta \vee_\theta [y]_\theta &:= [x \vee y]_\theta, \\
[x]_\theta \wedge_\theta [y]_\theta &:= [x \wedge y]_\theta, \\
\neg_\theta [x]_\theta &:= [\neg x]_\theta.
\end{aligned}
$$

The order $\leq$ on T_θ can be characterized as, $[x]_\theta \leq [y]_\theta \iff \neg y \leq \neg x$.

We next prove that the relation θ is compatible with the operators f and g.

Lemma 6.2. *Let $\langle T, g\rangle$ be a gKAN-algebra, and let $\theta \subseteq T \times T$ be the relation defined in 2. Then, the equivalence relation θ is compatible with the operations $\wedge$, $\vee$, $\neg$, as well as with the operators f and g.*

Proof. By Lemma 2.6, we know that θ is compatible with $\wedge$, $\vee$, and $\neg$. We now show that θ is also compatible with f and g. Let $(x, y) \in \theta$. Then $\neg x = \neg y$, which implies $f(\neg x) = f(\neg y)$. Using property (K4), we obtain $\neg g(x) = \neg g(y)$, hence $(g(x), g(y)) \in \theta$. Moreover, $g(\neg x) = g(\neg y)$, and by (K5) we have $\neg f(x) = \neg f(y)$, so $(f(x), f(y)) \in \theta$. □

Theorem 6.3. *Let $\langle T, g\rangle$ be a gKAN-algebra. Then $\langle T_\theta, f_\theta, g_\theta\rangle$ is a KGp-algebra, where the operators f_θ and g_θ are defined by:*

(i) $f_\theta([x]_\theta) = [f(x)]_\theta$,

(ii) $g_\theta([x]_\theta) = [g(x)]_\theta$.

Proof. By Lemma 6.2, the operators f_θ and g_θ are well-defined. We show that they form a Galois connection. Let $[x]_\theta, [y]_\theta \in T_\theta$, and suppose that $f_\theta([x]_\theta) \leq [y]_\theta$. Then $\neg y \leq \neg f(x)$. By (K12), it follows that $f(\neg y) \leq f(\neg f(x))$, and using (K4), (K5), and (K10), we obtain $\neg g(y) = f(\neg y) \leq f(\neg f(x)) = fg(\neg x) \leq \neg x$, which implies $[x]_\theta \leq g_\theta([y]_\theta)$. The converse is proved similarly. Moreover, each of the conditions (G1)(G5) holds directly by the corresponding axioms of the gKAN-algebra $\langle T, g\rangle$, namely: (G1) by (K1), (G2) by (K6), (G3) by (K14), (G4) by (K4), and (G5) by (K5). Therefore, $\langle T_\theta, f_\theta, g_\theta\rangle$ is a KGp-algebra. □

Example 6.4. *Consider the gKAN-algebra T from Example 6.1. We have:*

$$T_\theta = \{[0]_\theta, [d]_\theta, [e]_\theta, [h]_\theta, [1]_\theta\},$$

where $[0]_\theta = \{0, a, b, c\}$, $[d]_\theta = \{d\}$, $[e]_\theta = \{e\}$, $[h]_\theta = \{h\}$, $[1]_\theta = \{1\}$. Then, it can be verified that T_θ is a KGp-algebra whose Hasse diagram is shown in Figure 9, and whose operations are given in Table 8.

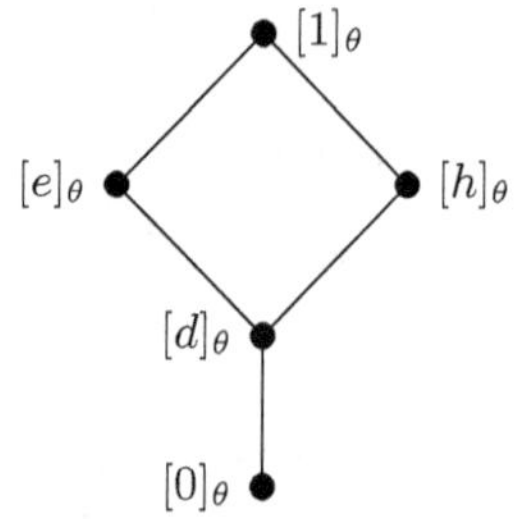

Figure 9

$[x]_\theta$	$\neg_\theta([x]_\theta)$	$f_\theta([x]_\theta)$	$g_\theta([x]_\theta)$
$[0]_\theta$	$[1]_\theta$	$[0]_\theta$	$[0]_\theta$
$[d]_\theta$	$[0]_\theta$	$[d]_\theta$	$[d]_\theta$
$[e]_\theta$	$[0]_\theta$	$[h]_\theta$	$[d]_\theta$
$[h]_\theta$	$[0]_\theta$	$[1]_\theta$	$[e]_\theta$
$[1]_\theta$	$[0]_\theta$	$[1]_\theta$	$[1]_\theta$

Table 8

We now show that both constructions induced by a gKAN-algebra are isomorphic to each other.

Theorem 6.4. *Let $\langle T, g\rangle$ be a gKAN-algebra. Then, the structures $\langle T_\theta, f_\theta, g_\theta\rangle$ and $\langle T^\Diamond, f^\Diamond, g^\Diamond\rangle$ are isomorphic.*

Proof. Let $\varphi : T_\theta \to T^\Diamond$ be the map defined by $\varphi([x]_\theta) := \Diamond x$. Then, by Proposition 2.3 (2), we have $\varphi([x]_\theta \vee_\theta [y]_\theta) = \Diamond(x \vee y) = \Diamond x \vee \Diamond y = \varphi([x]_\theta) \vee_\theta \varphi([y]_\theta)$ and $\varphi([x]_\theta \wedge_\theta [y]_\theta) = \Diamond(x \wedge y) = \Diamond(\Diamond x \wedge \Diamond y) = \Diamond(\varphi([x]_\theta) \wedge \varphi([y]_\theta)) = \varphi([x]_\theta) \wedge^\Diamond \varphi([y]_\theta)$. On the other hand, by Proposition 2.2 (8), we get $\varphi(\neg_\theta[x]_\theta) = \varphi([\neg x]_\theta) = \Diamond\neg x = \Diamond\neg\Diamond x = \neg^\Diamond\Diamond x = \neg^\Diamond\varphi([x]_\theta)$ that is, $\varphi(\neg_\theta[x]_\theta) = \neg^\Diamond\varphi([x]_\theta)$. Now, suppose that $[x]_\theta, [y]_\theta \in T_\theta$ and $\varphi([x]_\theta) = \varphi([y]_\theta)$. Then, $\Diamond x = \Diamond y$. By Proposition 2.2 (8), we have $\neg x = \neg y$, hence $[x]_\theta = [y]_\theta$. Moreover, for every $x \in T^\Diamond$ there exists $[x]_\theta \in T_\theta$ such that $\varphi([x]_\theta) = \Diamond x = x$. Therefore, the map $\varphi : T_\theta \to T^\Diamond$ given by $\varphi([x]) = \Diamond x$ is a isomorphism. It remains to check compatibility with g_θ and f_θ. By (K19), $\varphi(g_\theta([x])) = \varphi([g(x)]) = \Diamond g(x) = \Diamond g(\Diamond x) = g^\Diamond(\Diamond x) = g^\Diamond(\varphi([x]))$. Similarly, one can show that $\varphi(f_\theta([x])) = f^\Diamond(\varphi([x]))$. □

7 Kalman's Construction

Now, we present a Kalman-style construction for KGp-algebras. Using this construction, we prove that the category of KGp-algebras is equivalent to the category of gKANc-algebras.

Let $\langle A, f, g\rangle$ be a KGp-algebra, and consider $K(A) := \{(a,b) \in A\times A : a\wedge b = 0\}$. In [26], by defining the following operations on $K(A)$:

$$
\begin{aligned}
(a,b) \vee (x,y) &:= (a \vee x, b \wedge y),\\
(a,b) \wedge (x,y) &:= (a \wedge x, b \vee y),\\
\neg(a,b) &:= (a^*, a),\\
\sim (a,b) &:= (b,a),\\
0 &:= (0,1),\\
1 &:= (1,0),\\
c &:= (0,0),
\end{aligned}
$$

it follows that the algebra $K(A) = \langle K(A), \vee, \wedge, \sim, \neg, 0, 1, c\rangle$ is a $KANc$-algebra. Next, we define the following unary operators on $K(A)$:

$$
\begin{aligned}
g_K((a,b)) &:= (g(a), f(b)),\\
f_K((a,b)) &:= (f(a), g(b)).
\end{aligned}
$$

Lemma 7.1. *Let $\langle A, f, g\rangle$ be a KGp-algebra and let $(a,b) \in K(A)$. Then:*

1. $g_K(a,b) \in K(A)$,
2. $f_K(a,b) =\sim g_K(\sim (a,b))$,
3. $f_K(a,b) \in K(A)$.

Proof. We will only prove (1). The remaining properties follow directly from it. Since $(a,b) \in K(A)$, we have $a \wedge b = 0$. Then, by (G6) and (G2), it holds that $g(a) \wedge f(b) \le f(a \wedge b) = f(0) = 0$. Therefore, $g_K(a,b) \in K(A)$. □

Lemma 7.2. *Let $\langle A, f, g\rangle$ be a KGp-algebra. Then, $\langle K(A), g_K\rangle$ is a gKANc-algebra.*

Proof. By Lemma 7.1, the operator g_K is well-defined. We will prove axioms (K1) through (K6). Let $(a,b), (d,e) \in K(A)$.
(K1): From (G6), it follows that $g_K(1,0) = (g(1), f(0)) = (1,0)$.

(K2): From (G8) and (G7), we have

$$\begin{aligned} g_K((a,b)\wedge(d,e)) &= (g(a\wedge d), f(b\vee e)) \\ &= (g(a)\wedge g(d), f(b)\vee f(e)) \\ &= (g(a), f(b))\wedge(g(d), f(e)) \\ &= g_K(a,b)\wedge g_K(d,e). \end{aligned}$$

(K3): From (G9) and (G10), we get

$$\begin{aligned} (a,b)\wedge g_K(f_K(a,b)) &= (a,b)\wedge g_K(f(a), g(b)) \\ &= (a,b)\wedge(gf(a), fg(b)) \\ &= (a\wedge gf(a), b\vee fg(b)) \\ &= (a,b). \end{aligned}$$

That is, $(a,b)\leq g_K(f_K(a,b))$.

(K4): $f_K(\neg(a,b)) = f_K(a^*,a) = (f(a^*), g(a))$. Using (G4) and (G13), we have $f(a^*) = g(a)^*$, thus $f_K(\neg(a,b)) = (g(a)^*, g(a)) = \neg g_K(a,b)$.

(K5): $g_K(\neg(a,b)) = g_K(a^*,a) = (g(a^*), f(a))$. Using (G14), we have $g(a^*) = f(a)^*$, therefore $g_K(\neg(a,b)) = (f(a)^*, f(a)) = \neg f_K(a,b)$.

(K6): $g_K(a,b)\wedge f_K(c,d) = (g(a), f(b))\wedge(f(c), g(d)) = (g(a)\wedge f(c), f(b)\vee g(d))$. Using (G2) and (G3), we obtain $(g(a)\wedge f(c), f(b)\vee g(d)) \leq (f(a\wedge c), g(b\vee d)) = f_K(a\wedge c, b\vee d)$, hence $g_K(a,b)\wedge f_K(c,d)\leq f_K((a,b)\wedge(c,d))$.

Finally, we prove that $g_K(c) = c$. By (G6) and (G1), we have $g_K(0,0) = (g(0), f(0)) = (0,0)$. Therefore, $\langle K(A), g_K\rangle$ is a gKANc-algebra. □

Example 7.1. *Considering the KGp-algebra $\langle T^\Diamond, f^\Diamond, g^\Diamond\rangle$ from Example 6.3, we obtain the gKAN-algebra $\langle K(T^\Diamond), g_K^\Diamond\rangle$, represented by the Hasse diagram in Figure 10. The operations $\sim$, $\neg$, and $g_K^\Diamond$ are detailed in Table 9.*

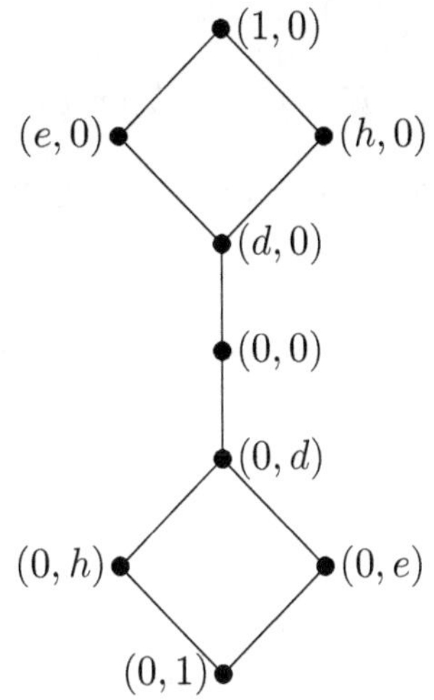

Figure 10

(x,y)	$\sim(x,y)$	$\neg(x,y)$	$g_K^{\Diamond}(x,y)$
$(0,0)$	$(0,0)$	$(1,0)$	$(0,0)$
$(0,d)$	$(d,0)$	$(1,0)$	$(0,d)$
$(0,e)$	$(e,0)$	$(1,0)$	$(0,h)$
$(0,h)$	$(h,0)$	$(1,0)$	$(0,1)$
$(0,1)$	$(1,0)$	$(1,0)$	$(0,1)$
$(d,0)$	$(0,d)$	$(0,d)$	$(d,0)$
$(e,0)$	$(0,e)$	$(0,e)$	$(h,0)$
$(h,0)$	$(0,h)$	$(0,h)$	$(1,0)$
$(1,0)$	$(0,1)$	$(0,1)$	$(1,0)$

Table 9

We denote by **KGp** the category whose objects are KGp-algebras, by **gKAN** the category whose objects are gKAN-algebras and by **gKANc** the category whose objects are gKANc-algebras. In both categories, morphisms are the respective algebra homomorphisms.

By Lemma 2.4 in [26], the construction K defines a functor from the category of distributive p-algebras to the category of KAN-algebras. Furthermore, given two

KGp-algebras $\langle A, f, g\rangle$ and $\langle B, f, g\rangle$, and a morphism $h : A \to B$ in **KGp**, the map

$$K(h) : K(A) \to K(B), \quad K(h)(x, y) = (h(x), h(y)),$$

is a morphism in **KANc**. Indeed, for every $(a, b) \in K(A)$, we have

$$\begin{aligned} K(h)(g_K(a, b)) &= K(h)(g(a), f(b)) \\ &= (h(g(a)), h(f(b))) \\ &= (g(h(a)), f(h(b))) \\ &= g_K(h(a), h(b)) \\ &= g_K(K(h)(a, b)). \end{aligned}$$

This shows that $K(h)$ preserves the operator g_K, and thus $K(h)$ is a morphism in **gKANc** from $\langle K(A), g_K\rangle$ to $\langle K(B), g_K\rangle$. Consequently, K is a functor from **KGp** to **gKANc**.

Furthermore, as shown in [26], if $A \in$ PDL and we consider $(a, b), (x, y) \in K(A)$, then

$$((a, b), (x, y)) \in \theta \quad \text{if and only if} \quad \neg(a, b) = \neg(x, y) \quad \text{if and only if} \quad a = x. \tag{3}$$

That is, two elements of $K(A)$ are θ-equivalent if and only if they agree on the first coordinate.

This fact allows us to define, for every gKAN-algebra T, a canonical map into the quotient KAN-algebra $K(T_\theta)$, which turns out to be a morphism in the category **gKAN**, as established below. This result will be used later.

Lemma 7.3. *Let $\langle T, g\rangle$ be a gKAN-algebra. Then, the map*

$$\rho : T \longrightarrow K(T_\theta), \quad \rho(x) = ([x]_\theta, [\sim x]_\theta),$$

*is an injective morphism in **gKAN**.*

Proof. By Lemma [26], ρ is an injective morphism in **KAN**. We verify that it preserves g. Indeed, $\rho(g(x)) = ([g(x)]_\theta, [\sim g(x)]_\theta)$ and $g_K(\rho(x)) = g_K([x]_\theta, [\sim x]_\theta) = (g_\theta([x]_\theta), f_\theta([\sim x]_\theta)) = ([g(x)]_\theta, [f(\sim x)]_\theta)$. Since $f(\sim x) =\sim g(x)$, it follows that $\rho(g(x)) = ([g(x)]_\theta, [\sim g(x)]_\theta) = ([g(x)]_\theta, [f(\sim x)]_\theta) = g_K(\rho(x))$. Thus, ρ preserves g, and we conclude that it is a morphism in **gKAN**. □

Let $T \in$ KANc. We consider the set $C(T) := \{x \in T : x \geq c\}$ and define the unary operation $\neg_c x := \neg x \vee c$ on $C(T)$.

Lemma 7.4. *Let $\langle T, g\rangle$ be a gKANc-algebra. Then, $\langle C(T), f, g\rangle$ is a KGp-algebra. Moreover, if S is a gKANc-algebra and $h : T \to S$ is a morphism in **gKANc**, then $C(h)$ is a morphism in **KGp**.*

Proof. According to the results in [26], the structure $\langle C(T), \vee, \wedge, \neg_c, c, 1\rangle$ is a p-algebra. We now show that $C(T)$ is closed under the operations f and g. Let $x \in C(T)$, i.e., $x \geq c$. By (K11), it follows that $g(x) \geq g(c) = c$, so $g(x) \in C(T)$. Since $\sim x \leq\sim c = c$, we have by (K11) that $g(\sim x) \leq g(c) = c$, and thus $c \leq\sim g(\sim x) = f(x)$. Hence, $f(x) \in C(T)$. By (K13), the pair (f, g) defines a Galois connection on $C(T)$. Axioms (G1) and (G2) follow directly from the definition of a gKANc-algebra, since (G1) is a consequence of (K1) and (G2) of (K6). Axiom (G3) is an immediate consequence of (K14). For (G4), by (K9) and (K4) we have $f(\neg_c x) = f(\neg x \vee c) = f(\neg x) \vee f(c) = \neg g(x) \vee c = \neg_c g(x)$. Similarly, (G5) holds because, by (K5), $g(\neg_c x) = g(\neg x \vee c) = g(\neg x) \vee g(c) = \neg f(x) \vee c = \neg_c f(x)$.

□

The following lemma establishes an isomorphism giving an alternative but equivalent algebraic structure to $C(T)$.

Lemma 7.5. *Let $\langle T, g\rangle$ be a gKANc-algebra. Then, the structures $\langle T^\Diamond, f^\Diamond, g^\Diamond\rangle$ and $\langle C(T), f, g\rangle$ are isomorphic.*

Proof. Let $h\colon T^\Diamond \to C(T)$ be the map defined by $h(x) := x \vee c$. Then, $h(0) = c$, $h(1) = 1$, and $h(x \vee y) = h(x) \vee h(y)$. By Lemma 2.4 (5), we have

$$h(x \wedge^\Diamond y) = (x \wedge^\Diamond y) \vee c = \Diamond(x \wedge y) \vee c = (x \wedge y) \vee c = (x \vee c) \wedge (y \vee c) = h(x) \wedge h(y).$$

Moreover, using Lemma 2.4 (5), it follows that

$$\neg_c h(x) = \neg(x \vee c) \vee c = (\neg x \wedge \neg c) \vee c = \neg x \vee c = \Diamond\neg x \vee c = \neg^\Diamond x \vee c = h(\neg^\Diamond x).$$

Next, we show that $h(g^\Diamond(x)) = g(h(x))$ and $h(f^\Diamond(x)) = f(h(x))$. Indeed, by (K17) and Lemma 6.1 (ii),

$$h(g^\Diamond(x)) = g^\Diamond(x) \vee c = \Diamond g(x) \vee c = g(\Diamond x) \vee c = g(x) \vee c = g(x \vee c) = g(h(x)).$$

Similarly, by (K18) and (K9),

$$h(f^\Diamond(x)) = f^\Diamond(x) \vee c = \Diamond f(x) \vee c = f(\Diamond x) \vee c = f(x) \vee c = f(x \vee c) = f(h(x)).$$

Finally, we prove that h is bijective. Let $x \in C(T)$. Consider $\Diamond x \in T^\Diamond$, which satisfies, by Lemma 2.4 (5), $x = x \vee c = \Diamond x \vee c = h(\Diamond x)$. Hence, h is surjective. On the other hand, let $x, y \in T^{\Diamond}$ such that $h(x) = h(y)$, i.e., $x \vee c = y \vee c$. Then, by Proposition 2.3 (2), we have $\Diamond x \vee \Diamond c = \Diamond y \vee \Diamond c$, and by Lemma 2.4 (2), it follows that $\Diamond x = \Diamond y$, which implies $x = y$. This proves that h is injective. Therefore, h is an isomorphism. □

Corollary 7.1. *Let $\langle T, g\rangle$ be a gKANc-algebra. Then, the algebras $\langle K(C(T)), g_K\rangle$ and $\langle K(T^{\Diamond}), g_K\rangle$ are isomorphic.*

Proof. By Lemma 7.5, the map $h : T^{\Diamond} \to C(T)$ defined by $h(x) = x \vee c$ is an isomorphism. Then, the map $K(h) : K(T^{\Diamond}) \to K(C(T))$ given by $K(h)(a, b) = (h(a), h(b))$ for all $(a, b) \in K(T^{\Diamond})$, is the desired isomorphism. □

Let T and S be gKANc-algebras, and let $h : T \to S$ be a homomorphism. By Lemma 7.4, the assignments $T \mapsto C(T)$ and $h \mapsto C(h)$ define a functor C from the category **gKANc** to the category **KGp**.

Lemma 7.6. *Let $\langle A, f, g\rangle$ be a KGp-algebra. Then, the map $\alpha : A \to C(K(A))$ defined by $\alpha(x) = (x, 0)$ is an isomorphism in the category* **KGp**.

Proof. We show that α commutes with the unary operators f and g. Let $x \in A$. Then, by (G6) and (G1), we have $\alpha(g(x)) = (g(x), 0) = (g(x), f(0)) = g_K((x, 0)) = g_K(\alpha(x))$. Similarly, $\alpha(f(x)) = (f(x), 0) = (f(x), g(0)) = f_K((x, 0)) = f_K(\alpha(x))$. This proves that α is a morphism in **KGp**. Since α is clearly bijective, we conclude that it is an isomorphism. □

Let $\langle A, f, g\rangle$ be a KGp-algebra. Consider the subset of $K(A)$ defined by

$$K(A)^{\Diamond} := \{(a, b) \in K(A) :\sim (a, b) = \neg(a, b)\}.$$

Since $\sim (a, b) = (b, a)$ and $\neg(a, b) = (a^*, a)$, it follows that $b = a^*$. Therefore,

$$K(A)^{\Diamond} = \{(a, a^*) : a \in A\}.$$

The operations on $K(A)^{\Diamond}$ are defined as follows:

$$\begin{aligned}
(a, a^*) \vee^{\Diamond} (b, b^*) &:= (a \vee b, (a \vee b)^*), \\
(a, a^*) \wedge^{\Diamond} (b, b^*) &:= (a \wedge b, (a \wedge b)^*), \\
\neg^{\Diamond}(a, a^*) &:= (a^*, a^{**}), \\
f_K^{\Diamond}(a, a^*) &:= (f(a), f(a)^*), \\
g_K^{\Diamond}(a, a^*) &:= (g(a), g(a)^*), \\
0 &:= (0, 1), \\
1 &:= (1, 0).
\end{aligned}$$

As a direct consequence of Lemma 7.6 and Lemma 7.5, we obtain the following characterization of any KGp-algebra.

Corollary 7.2. *Let $\langle A, f, g\rangle$ be a KGp-algebra. Then, the structures $\langle A, f, g\rangle$ and $\langle K(A)^{\Diamond}, f_K^{\Diamond}, g_K^{\Diamond}\rangle$ are isomorphic.*

Lemma 7.7. *Let $\langle T, g\rangle$ be a gKANc-algebra. Then, the map*

$$\beta : T \to K(C(T)), \quad \beta(x) := (x \vee c, \sim x \vee c)$$

is an isomorphism in the category **gKANc**.

Proof. It suffices to show that β commutes with the unary operator g. Let $x \in T$. Since $g(c) = c$, it follows that $f(c) = c$. Hence, using axiom (K9), we have:

$$\begin{aligned}\beta(g(x)) &= (g(x) \vee c, \ \sim g(x) \vee c)\\ &= (g(x) \vee c, \ f(\sim x) \vee f(c))\\ &= (g(x \vee c), \ f(\sim x \vee c))\\ &= g_K(\beta(x)).\end{aligned}$$

This proves that β is a morphism in **gKANc**. The remaining properties (injectivity, surjectivity, and preservation of other operations) follow from the context and previous results, so β is an isomorphism. □

From Lemmas 7.6 and 7.7, the following theorem can be deduced.

Theorem 7.3. *The functors K and C establish a categorical equivalence between the categories* **KGp** *and* **gKANc**, *accompanied by the natural isomorphisms α and β.*

8 Congruences on $K(A)$

Let $A \in$ PDL and let R be a congruence on A. Then, the relation S_R defined by

$$((a, b), (d, e)) \in S_R \quad \text{if and only if} \quad (a, d) \in R \text{ and } (b, e) \in R$$

is a congruence on $K(A)$. If S is a congruence on $K(A)$, then the relation R^S defined by

$$(a, b) \in R^S \quad \text{if and only if} \quad ((a, 0), (b, 0)) \in S$$

is a congruence on A.

In [26], it was shown that if $A \in$ PDL, the assignments $R \mapsto S_R$ and $S \mapsto R^S$ define an order isomorphism between $\mathrm{Con}(A)$ and $\mathrm{Con}(K(A))$. This result can be extended to KGp-algebras.

Lemma 8.1. *Let $\langle A, f, g\rangle$ be a KGp-algebra. Then:*

1. *If $R \in \mathrm{Con}(A)$, then $S_R \in \mathrm{Con}(K(A))$.*
2. *If $S \in \mathrm{Con}(K(A))$, then $R^S \in \mathrm{Con}(A)$.*

Proof. We prove (1) by showing that S_R is compatible with the operator g_K. Compatibility means that whenever $(a, b)\, S_R\, (x, y)$, it follows that $g_K(a, b)\, S_R\, g_K(x, y)$. Assume $(a, b)\, S_R\, (x, y)$. By definition of S_R, this implies $(a, x) \in R$ and $(b, y) \in R$. Since R is a congruence on A and f, g are operators on A, it follows that $(g(a), g(x)) \in R$ and $(f(b), f(y)) \in R$. Hence, by the definition of S_R, we have $g_K(a, b) = (g(a), f(b))\, S_R\, (g(x), f(y)) = g_K(x, y)$, so S_R is compatible with g_K. Therefore, S_R is a congruence on $K(A)$.

For (2), suppose $(a, b) \in R^S$. By definition of R^S, this means $(a, 0)\, S\, (b, 0)$. Since S is a congruence on $K(A)$ and is compatible with g_K, we have $g_K(a, 0) = (g(a), f(0))\, S\, (g(b), f(0)) = g_K(b, 0)$, which implies $(g(a), g(b)) \in R^S$. Furthermore, since S is compatible with $\sim$ and g_K, we deduce that $f_K(a, 0) =\sim g_K(\sim 0, \sim a) =\sim g_K(1, \sim a) = (f(a), g(0))\, S\, (f(b), g(0)) = f_K(b, 0)$, and therefore $(f(a), f(b)) \in R^S$. This shows that R^S is compatible with both f and g, and hence R^S is a congruence on A. □

The following result is an immediate consequence of Lemma 8.1.

Theorem 8.1. *Let $\langle A, f, g\rangle$ be a KGp-algebra. Then the assignments $R \mapsto S_R$ and $S \mapsto R^S$ define an order isomorphism between the ordered sets $\mathrm{Con}(A)$ and $\mathrm{Con}(K(A))$.*

9 Sendlewski's Construction

Let $A \in \mathrm{PDL}$ and $S \subseteq A$ a Boolean filter. We consider the set

$$K(A, S) := \{(a, b) \in A \times A : a \wedge b = 0 \text{ and } a \vee b \in S\},$$

which will be the underlying set of the algebraic structure constructed from A and S.

Theorem 9.1. *Let $\langle A, f, g\rangle$ be a KGp-algebra and S a KGp-Boolean filter of A. Then, $\langle K(A, S), g_K\rangle$ is a gKAN-algebra.*

Proof. By [26, Theorem 2.11], $K(A, S)$ is a subalgebra of $K(A)$, so it suffices to prove closure under g_K. Let $(a, b) \in K(A, S)$. Then, $g_K(a, b) = (g(a), f(b))$. Using (K6) and (K8), we have $g(a) \wedge f(b) \leq f(a \wedge b) = 0$. Since S is a KGp-filter, (K14) implies $g(a) \vee f(b) \in S$, and thus $g_K(a, b) \in K(A, S)$. □

Example 9.1. *Let $\langle T, g\rangle$ the KGp-algebra from Example 6.3 and the KGp-Boolean filter $S = \{d, e, h, 1\}$, we obtain the gKAN-algebra $(K(T^{\Diamond}, S), g_K^{\Diamond})$ whose Hasse diagram is shown in Figure 11. The operator g_K is given in Table 10.*

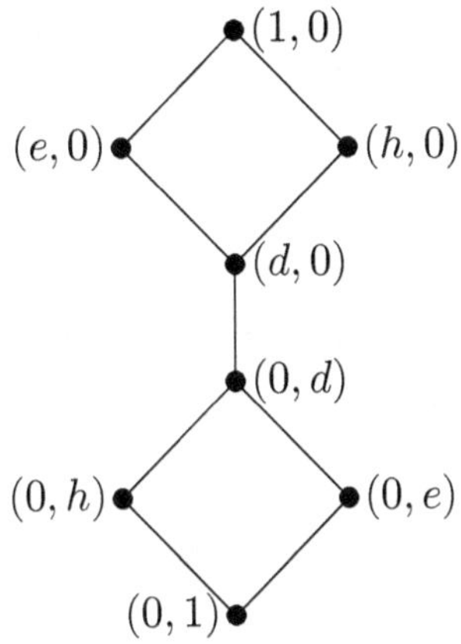

Figure 11

(x, y)	$g_K^{\Diamond}(x, y)$
$(0, 1)$	$(0, 1)$
$(0, h)$	$(0, 1)$
$(0, e)$	$(0, h)$
$(0, d)$	$(0, d)$
$(d, 0)$	$(d, 0)$
$(e, 0)$	$(d, 0)$
$(h, 0)$	$(e, 0)$
$(1, 0)$	$(1, 0)$

Table 10

Theorem 9.2. *Let A and A' be KGp-algebras, and let S and S' be KGp-Boolean filters of A and A', respectively. If $r : A \to A'$ is a morphism of KGp-algebras such that $r(S) \subseteq S'$, then*

$$K(r) : K(A, S) \to K(A', S'), \quad K(r)(a, b) = (r(a), r(b)),$$

is a morphism of gKAN-algebras.

Proof. By [26], $K(r)$ is a morphism of KAN-algebras. Since Theorem 9.1 establishes that $K(A,S)$ is a gKAN-algebra, it remains to show that $K(r)$ preserves the operator g_K. Indeed, we have $K(r)(g_K(a,b)) = K(r)(g(a), f(b)) = (r(g(a)), r(f(b)))$, and because r is a morphism of KGp-algebras, it preserves both f and g. Therefore, $(r(g(a)), r(f(b))) = (g(r(a)), f(r(b))) = g_K(r(a), r(b)) = g_K(K(r)(a,b))$, which shows that $K(r)$ preserves g_K, as required. □

The following lemma shows that, starting from the Sendlewski construction applied to a KGp-algebra and a suitable Boolean filter, the original algebra can be recovered as a quotient. This confirms that the construction faithfully reflects the structure of the original algebra.

Lemma 9.1. *Let $\langle A, f, g\rangle$ be a KGp-algebra and let S be a KGp-Boolean filter. Then, the map*

$$h : K(A,S)/\theta \longrightarrow A, \quad h([(a,b)]_\theta) = a,$$

is an isomorphism of KGp-algebras.

Proof. It is clear that h is well-defined. If $h([(a,b)]_\theta) = h([(c,d)]_\theta)$, then $a = c$, which, by 3, implies $[(a,b)]_\theta = [(c,d)]_\theta$, and thus h is injective. For each $a \in A$, we have $(a,1) \in K(A,S)$ and $h([(a,1)]_\theta) = a$, so h is surjective. Moreover, h preserves the operators f_θ and g_θ, since $h(g_\theta([(a,b)]_\theta)) = h([(g(a), f(b))]_\theta) = g(a)$ and $g(h([(a,b)]_\theta)) = g(a)$, and similarly for f_θ. Therefore, h is an isomorphism. □

Lemma 9.2. *Let $\langle T, g\rangle$ be a gKAN-algebra. Then,*

$$T^+ := \{x \in T : \Box x = 1\}$$

is a filter. Moreover, T^+/θ is a KGp-Boolean filter of T_θ.

Proof. According to [26, Lemma 2.14], it suffices to prove that T^+/θ is closed under g. Suppose $x \in T^+$. Then, $\Box x = 1$, hence $g(\Box x) = 1$. By (K15), it follows that $\Box g(x) = 1$, so $g(x) \in T^+$. Therefore, $g([x]_\theta) \in T^+/\theta$, and thus $T^+/\theta \subseteq T/\theta$ is a KGp-Boolean filter. □

Example 9.2. *The KGp-algebra $\langle T_\theta, f_\theta, g_\theta\rangle$ from Example 6.4 has*

$$T^+/\theta = \{[d], [e], [f], [1]\}$$

as one of its KGp-Boolean filters.

Example 9.3. *From the KGp-Boolean filter given in Example 9.2, we obtain the gKAN-algebra $\langle K(T_\theta, T^+/\theta), g_K\rangle$, whose Hasse diagram is shown in Figure 12. The operator g_K is presented in Table 11.*

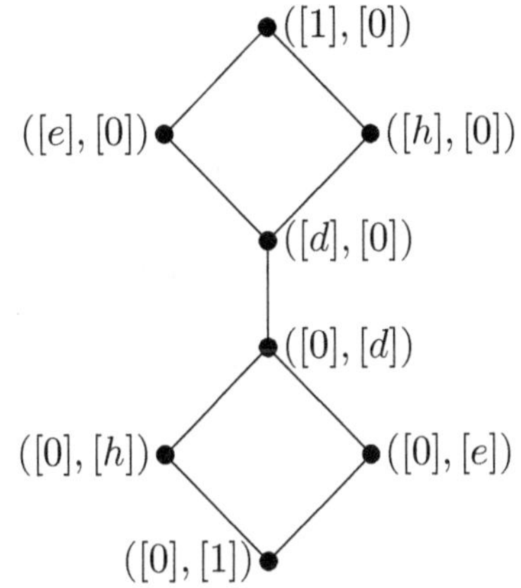

Figure 12

$([x],[y])$	$g_K([x],[y])$
$([0],[1])$	$([0],[1])$
$([0],[h])$	$([0],[1])$
$([0],[e])$	$([0],[h])$
$([0],[d])$	$([0],[d])$
$([d],[0])$	$([d],[0])$
$([e],[0])$	$([d],[0])$
$([h],[0])$	$([e],[0])$
$([1],[0])$	$([1],[0])$

Table 11

We now prove that every gKAN-algebra is isomorphic to a gKAN-algebra obtained through the Sendlewski construction applied to its quotient algebra and the corresponding Boolean filter.

Theorem 9.3. *Let $\langle T, g\rangle$ be a gKAN-algebra. Then, $\langle T, g\rangle$ and $\langle K(T_\theta, T^+/\theta), g_K\rangle$ are isomorphic.*

Proof. By [26, Theorem 2.15], the map

$$\rho : T \to K(T/\theta, T^+/\theta), \quad \rho(x) = ([x]_\theta, [\sim x]_\theta),$$

is an isomorphism in **KAN**. Moreover, as shown in the proof of Lemma 7.3, one can verify that ρ preserves g. Therefore, ρ is an isomorphism in **gKAN**. □

We define the category **KGpS** whose objects are pairs (A, S), where A is a KGp-algebra and S is a KGp-Boolean filter of A, and whose morphisms $r : (A, S) \to (A', S')$ are morphisms in **KGp**, $r : A \to A'$, such that $r(S) \subseteq S'$.

Theorem 9.4. *Let $\langle T, g\rangle$ and $\langle T', g'\rangle$ be two gKAN-algebras, and let $r : T \to T'$ be a morphism in **gKAN**. Then,*

$$r_\theta : T_\theta \to T'/\theta, \quad r_\theta([x]_\theta) = [r(x)]_\theta,$$

*is a morphism in **KGp**, and $r_\theta(T^+/\theta) \subseteq T'^+/\theta$.*

Proof. By [26, Theorem 2.15], it remains to prove that r_θ preserves the operators f_θ and g_θ. For g_θ, note that $r_\theta(g_\theta([x]_\theta)) = r_\theta([g(x)]_\theta) = [r(g(x))]_\theta$. Since r is a morphism in **gKAN**, we have $[r(g(x))]_\theta = [g(r(x))]_\theta = g_\theta([r(x)]_\theta) = g_\theta(r_\theta([x]_\theta))$, hence $r_\theta(g_\theta([x]_\theta)) = g_\theta(r_\theta([x]_\theta))$. The case for f_θ is verified similarly. □

Taking into account the results obtained in this section, the following theorem can be proved.

Theorem 9.5. *The functor K establishes a categorical equivalence between the category **gKAN** and the category **KGpS**.*

10 Monteiros Construction

Given a Nelson algebra, it is always possible to obtain a centered Nelson algebra through a process known as Monteiros construction, which, notably, was never published. In [7], this construction served as inspiration to derive a centered Stone KAN-algebra from a Stone KAN-algebra. In this section, we build upon these ideas to construct a gKANc-algebra from a gKAN-algebra.

Let $\langle T, g\rangle$ be a gKAN-algebra. Consider the set

$$M(T) := \{(a, b) \in T^\Diamond \times T^\Box : a \le b\}.$$

In [7], by defining the following operations on $M(T)$:

$$\begin{aligned}
(a,b) \cup (d,e) &:= (a \vee d,\ \Box(b \vee e)),\\
(a,b) \cap (d,e) &:= (\Diamond(a \wedge d),\ b \wedge e),\\
\boxplus(a,b) &:= (\sim b,\ \sim a),\\
\circledast(a,b) &:= (\Diamond\neg a,\ \sim a),\\
0_M &:= (0,\ 0),\\
1_M &:= (1,\ 1),\\
c_M &:= (0,\ 1),
\end{aligned}$$

it follows that the algebra $M(T) = \langle M(T), \cup, \cap, \boxplus, \circledast, c_M, 0_M, 1_M\rangle$ is a KANc-algebra. Moreover, if T is a KANc-algebra, then T is isomorphic to $M(T)$. Here, the order $\subseteq$ on $M(T)$ is given by: $(a,b) \subseteq (d,e)$ if and only if $a \leq d$ and $b \leq e$.

We now show that Monteiros construction can be extended to the framework of gKAN-algebras. We define the following unary operators on $M(T)$:

$$g_M(a,b) := (\Diamond g(a), \Box g(b))$$

and

$$f_M(a,b) := (\Diamond f(a), \Box f(b)).$$

Lemma 10.1. *Let $\langle T, g\rangle$ be a gKAN-algebra and let $(a,b) \in M(T)$. Then:*

1. *$g_M(a,b) \in M(T)$,*
2. *$f_M(a,b) = \boxplus g_M(\boxplus(a,b))$,*
3. *$f_M(a,b) \in M(T)$.*

Proof. (1): Since $(a,b) \in M(T)$, we have $a \in T^{\Diamond}$, $b \in T^{\Box}$, and $a \leq b$. By (K11), it follows that $g(a) \leq g(b)$. Moreover, by Propositions 2.3 and 2.2, we have $\Diamond g(a) \in T^{\Diamond}$, $\Box g(b) \in T^{\Box}$, and $\Diamond g(a) \leq g(a) \leq g(b) \leq \Box g(b)$. Therefore, $g_M(a,b) \in M(T)$.

(2): Using (K23) and (K24), we compute:

$$\begin{aligned}
\boxplus g_M(\boxplus(a,b)) &= \boxplus g_M(\sim b, \sim a)\\
&= \boxplus(\Diamond g(\sim b), \Box g(\sim a))\\
&= (\sim \Box g(\sim a), \sim \Diamond g(\sim b))\\
&= (\Diamond f(a), \Box f(b))\\
&= f_M(a,b).
\end{aligned}$$

(3): This follows from (1), (2), and the fact that $\boxplus$ is closed in $M(T)$. $\Box$

Theorem 10.1. *Let $\langle T, g\rangle$ be a gKAN-algebra. Then, $\langle M(T), g_M\rangle$ is a gKANc-algebra.*

Proof. By Lemma 10.1, the operator g_M is well defined. Let $(a, b), (d, e) \in M(T)$. We verify axioms (K1) through (K6).
(K1): By Proposition 2.2 (5) and (K1), we have $g_M(0, 0) = (\Diamond g(0), \Box g(0)) = (\Diamond 0, \Box 0) = (0, 0)$.
(K2): By Proposition 2.3 (1) and (K2)

$$\begin{aligned} g_M((a,b)\cap(d,e)) &= g_M(\Diamond(a\wedge d), b\wedge e) \\ &= (\Diamond\Diamond g(a\wedge d), \Box g(b\wedge e)) \\ &= (\Diamond g(a\wedge d), \Box g(b\wedge e)) \\ &= (\Diamond(g(a)\wedge g(d)), \Box(g(b)\wedge g(e))) \\ &= (\Diamond(\Diamond g(a)\wedge\Diamond g(d)), \Box g(b)\wedge\Box g(e)) \\ &= (\Diamond g(a), \Box g(b))\cap(\Diamond g(d), \Box g(e)) \\ &= g_M(a,b)\cap g_M(d,e). \end{aligned}$$

(K3): Since $(a, b) \in M(T)$, we have $\Diamond a = a$, $\Box b = b$, and $a \leq b$. We show that $(a, b) \subseteq g_M f_M(a, b)$.

We compute $g_M f_M(a, b) = g_M(\Diamond f(a), \Box f(b)) = (\Diamond g(\Diamond f(a)),\ \Box g(\Box f(b)))$. On the other hand, by (K3), Proposition 2.2, and (K19), we obtain $a = \Diamond a \leq \Diamond g f(a) = \Diamond g(\Diamond f(a))$. Similarly, we can prove $b \leq \Box g(\Box f(b))$. Therefore, $(a, b) \subseteq g_M f_M(a, b)$.
(K4): Follows from Proposition 2.2 (8) and (K4)

$$\begin{aligned} \circledast g_M(a,b) &= \circledast(\Diamond g(a), \Box g(b)) \\ &= (\Diamond\neg\Diamond g(a), \sim\Diamond g(a)) \\ &= (\Diamond\neg g(a), \neg g(a)) \\ &= (\Diamond f(\neg a), \Box\neg g(a)) \\ &= (\Diamond f(\Diamond\neg a), \Box f(\sim a)) \\ &= f_M(\Diamond\neg a, \sim a) \\ &= f_M(\circledast(a,b)). \end{aligned}$$

(K5): A similar argument as in (K4) applies.
(K6): Follows from Proposition 2.3 (3) and (K6)

$$\begin{aligned}
g_M(a,b) \cap f_M(d,e) &= (\Diamond g(a), \Box g(b)) \cap (\Diamond f(d), \Box f(e)) \\
&= (\Diamond(\Diamond g(a) \wedge \Diamond f(d)), \Box g(b) \wedge \Box f(e)) \\
&= (\Diamond(g(a) \wedge f(d)), \Box(g(b) \wedge f(e))) \\
&\subseteq (\Diamond f(a \wedge d), \Box f(b \wedge e)) \\
&= (\Diamond f(\Diamond(a \wedge d)), \Box f(b \wedge e)) \\
&= f_M(\Diamond(a \wedge d), b \wedge e) \\
&= f_M((a,b) \cap (d,e)).
\end{aligned}$$

Finally, we verify $g_M(c) = c$. By (K1) and Proposition 2.2 (4) y (5), we have $g_M(0,1) = (\Diamond g(0), \Box g(1)) = (\Diamond 0, \Box 1) = (0,1)$. Therefore, $\langle M(T), g_M\rangle$ is a gKANc-algebra. □

Example 10.1. *Let us consider the gKAN algebra T from Example 6.1. We obtain the gKAN-algebra $\langle M(T), g_K\rangle$, represented by the Hasse diagram in Figure 13. The operations $\sim$, $\neg$, and the operator g_K are detailed in Table 12.*

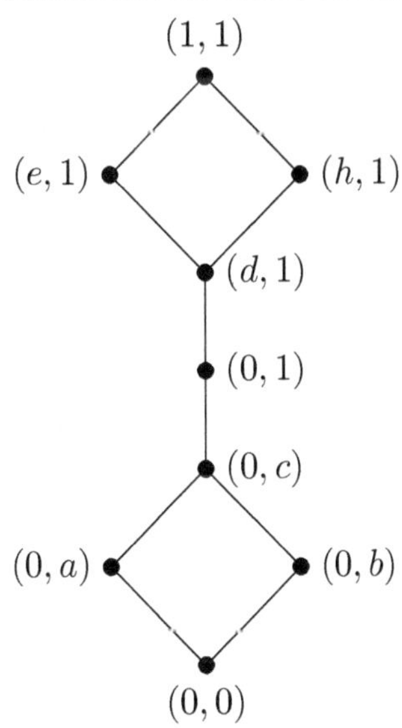

Figure 13

(x,y)	$\boxplus(x,y)$	$\circledast(x,y)$	$g_M(x,y)$
$(0,0)$	$(1,1)$	$(1,1)$	$(0,0)$
$(0,a)$	$(h,1)$	$(1,h)$	$(0,0)$
$(0,b)$	$(e,1)$	$(1,e)$	$(0,a)$
$(0,c)$	$(d,1)$	$(1,d)$	$(0,c)$
$(0,1)$	$(0,1)$	$(1,0)$	$(0,1)$
$(d,1)$	$(0,c)$	$(0,0)$	$(d,1)$
$(e,1)$	$(0,b)$	$(0,0)$	$(d,1)$
$(h,1)$	$(0,a)$	$(0,0)$	$(e,1)$
$(1,1)$	$(0,0)$	$(0,0)$	$(1,1)$

Table 12

Theorem 10.2. *Let $\langle T, g\rangle$ be a gKAN-algebra. Then, $\langle T, g\rangle$ is isomorphic to a subalgebra of a gKANc-algebra.*

Proof. By Theorem 10.1, the structure $\langle M(T), g_M\rangle$ is a gKANc-algebra. Consider the map $\delta : T \to M(T)$ defined by $\delta(x) = (\Diamond x, \Box x)$, which, according to [7, Theorem 4.2], is an injective homomorphism of KANc-algebras. Then, using (K19) and (K21), we obtain $\delta(g(a)) = (\Diamond g(a), \Box g(a)) = (\Diamond g(\Diamond a), \Box g(\Box a)) = g_M(\Diamond a, \Box a) = g_M(\delta(a))$, as desired. $\Box$

Theorem 10.3. *Let $\langle T, g\rangle$ be a gKAN-algebra. Then, the structures $\langle M(T), g_M\rangle$ and $\langle K(T^\Diamond), g_K^\Diamond\rangle$ are isomorphic.*

Proof. By Theorem 6.2, the structure $\langle T^\Diamond, f^\Diamond, g^\Diamond\rangle$ is a KGp-algebra. Moreover, by Lemma 7.2, $\langle K(T^\Diamond), g_K^\Diamond\rangle$ is a gKANc-algebra. According to , the map $t : M(T) \to K(T^\Diamond)$ defined by $t(x,y) = (x, \sim y)$ is an isomorphism of KANc-algebras. From (K23), it follows that $\sim \Box g(x) = \Diamond f(\sim x)$.

Thus, we have:

$$\begin{aligned} t(g_M(a,b)) &= t(\Diamond g(a), \Box g(b)) \\ &= (\Diamond g(a), \sim \Box g(b)) \\ &= (\Diamond g(a), \Diamond f(\sim b)) \\ &= (\Diamond g(a), f^\Diamond(\sim b)) \\ &= g_K^\Diamond(a, \sim b) \\ &= g_K^\Diamond(t(a,b)), \end{aligned}$$

that is, $t(g_M(a,b)) = g_K^\Diamond(t(a,b))$. Therefore, $\langle M(T), g_M\rangle$ and $\langle K(T^\Diamond), g_K^\Diamond\rangle$ are isomorphic. $\Box$

From Lemmas 7.5 and 7.7 and Theorem 10.3, we obtain the following result.

Corollary 10.4. *Let* $\langle T, g\rangle$ *be a gKANc-algebra. Then, the gKANc-algebras* $\langle T, g\rangle$ *and* $\langle M(T), g_M\rangle$ *are isomorphic.*

Acknowledgments

The author wishes to thank the anonymous referees for their meticulous work and constructive comments that significantly improved this manuscript. Special gratitude is extended to Professors Sergio Celani and Gustavo Pelaitay for their support, fruitful discussions, and suggestions that were essential for the completion of this work.

References

[1] Balbes R., Dwinger P: *Distributive lattices.* University of Missouri Press, 1974.

[2] Bezhanishvili, Guram. Glivenko type theorems for intuitionistic modal logics. Studia Logica 67 (2001), no. 1, 89109.

[3] Bimbo K., Dunn J.M.: *Generalized Galois logics. Relational semantics of nonclassical logical calculi.* CSLI Lecture Notes, Stanford, 2008.

[4] Blyth T.S.: *Lattices and ordered algebraic structures.* Springer, London, 2005.

[5] Boicescu, V.; Filipoiu, A.; Georgescu, G.; Rudeanu, S.: *ukasiewicz-Moisil algebras.* Ann. Discrete Math. **49**, North-Holland Publishing Co., Amsterdam, 1991, xvi+583 pp.

[6] Castaño D., Castaño V., Díaz Varela J. P., Muñoz Santis M.: *Varieties of pseudocomplemented Kleene algebras.* MLQ Math. Log. Q. **67** (2021), no. 1, 88–104.

[7] Calomino, I.; Pelaitay, G. (2025). *A new categorical equivalence for Stone algebras.* Mathematica Slovaca, 75, 1275–1292.

[8] Calomino, I., Pelaitay, G., Valverde, L.F.: *Heyting Algebras with Kalman-Galois Connections.* Stud. Logica (2025) `10.1007/s11225-025-10179-9`.

[9] Calomino, I.; Lewis-Smith, A.; Pelaitay, G.A. (2025). *Monadic pseudocomplemented distributive lattices.* Journal of Applied Logic, ISSN 1570-8683.

[10] Celani S., Jansana R.: *A note on the model theory for positive modal logic.* Fund. Inform. **114** (2012), no. 1, 31–54.

[11] Celani S., Jansana R.: *Priestley duality, a Sahlqvist theorem and a Goldblatt-Thomason theorem for positive modal logic.* Log. J. IGPL **7** (1999), no. 6, 683–715.

[12] Celani S., Jansana R.: *A new semantics for positive modal logic.* Notre Dame J. Formal Logic **38** (1997), no. 1, 1–18.

[13] Celani S., Montangie D.: *Hilbert algebras with Hilbert-Galois connections.* Stud. Log. **111** (2023), 113–138.

[14] Chajda I., Halaš R., Kühr J.: *Semilattice structures.* Research and Exposition in Mathematics, Heldermann Verlag, 2007.

[15] Chajda I., Paseka J.: *Galois connections and tense operators on q-effect algebras.* Fuzzy Sets Syst. **298** (2016), 56–68.

[16] Cignoli, R., Torrens Torrell, A. (2004). *Glivenko like theorems in natural expansions of BCK-logic.* MLQ Math. Log. Q, 50(2), 111125.

[17] Davey B., Priestley H.: *Introduction to lattice and order.* Cambridge University Press, 2002.

[18] Denecke K., Erné M., Wismath S.L.: *Galois connections and applications.* Mathematics and its Applications, Kluwer Academic Publishers, 2004.

[19] Dzik W., Järvinen J., Kondo M.: *Intuitionistic propositional logic with Galois connections.* Log. J. IGPL **18** (2010), 837–858.

[20] Dzik W., Järvinen J., Kondo M.: *Intuitionistic modal logic with a Galois connection has the finite model property.* Log. J. IGPL **21** (2013), 199–204.

[21] Dzik W., Järvinen J., Kondo M.: *Characterizing intermediate tense logics in terms of Galois connections.* Log. J. IGPL **22** (2014), 992–1018.

[22] Dzik W., Järvinen J., Kondo M.: *Representing expansions of bounded distributive lattices with Galois connections in terms of rough sets.* Int. J. Approx. Reasoning **55** (2014), 427–435.

[23] Figallo A. V., Pelaitay G., Sarmiento J.: *An algebraic study of tense operators on Nelson algebras.* Studia Logica **109** (2021), no. 2, 285–312.

[24] Figallo A. V., Sarmiento J., Figallo M.: *Tense distributive lattices: algebra, logic and topology.* Preprint, arXiv:2309.04771 [math.LO] (2023).

[25] Glivenko, V.: *Sur quelques points de la logique de M. Brouwer.* Academie Royale de Belgique, Bulletins de la classe des sciences, 15(5) (1929), 183–188.

[26] Gomez, C., Marcos, M. A., San Martín, H. J.: *On the relation of negations in Nelson algebras.* Rep. Math. Logic No, 56 (2021), 15–56.

[27] Jankowski A.W.: *Galois structures.* Stud. Log. **44** (1985), 109–124.

[28] Järvinen J., Radeleczki S.: *Representing regular pseudocomplemented Kleene algebras by tolerance-based rough sets.* J. Aust. Math. Soc. **105** (2018), no. 1, 57–78.

[29] Järvinen J., Radeleczki S.: *Prime filter structures of pseudocomplemented Kleene algebras and representation by rough sets.* Acta Math. Hungar. **160** (2020), no. 1, 175–196.

[30] Järvinen, J.; Kondo, M.; Kortelainen, J.: *Modal-like operators in Boolean lattices, Galois connections and fixed points.* Fundam. Inform. **76** (2007), 129–145.

[31] Kalman J.A.: *Lattices with involution.* Trans. Am. Math. Soc. **87** (1958), 485–491.

[32] Katriák, T.: *The structure of distributive double p-algebras. Regularity and congruences.* Algebra Univers. **3** (1973), 238–246.

[33] Monteiro, A. A.: *Sur les algèbres de Heyting symétriques.* (French) [Symmetric Heyting algebras] Special issue in honor of António Monteiro. Portugal. Math. **39** (1980), no. 1-4, 1–237 (1985).

[34] Orłowska E., Rewitzky I.: *Algebras for Galois-style connections and their discrete duality.* Fuzzy Sets Syst. **161** (2010), 1325–1342.

[35] Ore O.: *Galois connexions.* Trans. Am. Math. Soc. **55** (1944), 493–513.

[36] Pelaitay G., Zuluaga W.: *Tense operators on distributive lattices with implication.* Studia Logica **111** (2023), no. 4, 687–708.

[37] Sankappanavar H.P.: *Semi-Heyting algebras: An abstraction from Heyting algebras.* Actas del IX Congreso Dr. Antonio A. R. Monteiro (2007), 33–66.

[38] Sankappanavar H. P.: *Pseudocomplemented Ockham and De Morgan algebras.* MLQ Math. Log. Q. **32** (1986), 385–394.

[39] von Karger, B.: *Temporal algebra.* Math. Struct. Comput. Sci. **8** (1995), 277–320.

[40] Wang J., Wang M., Pelaitay G.: *An algebraic study of t-norm based residuated fuzzy tense logics.* Fuzzy Sets and Systems **520** (2025), Paper No. 109547, 28 pp.

[41] Wang J. T., She Y. H., He P. F., Ma N. N.: *On categorical equivalence of weak monadic residuated distributive lattices and weak monadic c-differential residuated distributive lattices.* Studia Logica **111** (2023), no. 3, 361–390.

Received August 2025

www.ingramcontent.com/pod-product-compliance
Lightning Source LLC
LaVergne TN
LVHW082343200926
848481LV00012B/3433
* 9 7 8 1 8 4 8 9 0 5 0 8 5 *